# *Resource Exploitation in Native North America*

**Recent Titles in**
**Native America: Yesterday and Today**
*Bruce E. Johansen, Series Editor*

George Washington's War on Native America
*Barbara Alice Mann*

The Native Peoples of North America: A History
Two Volumes
*Bruce E. Johansen*

Daughters of Mother Earth: The Wisdom of Native American Women
*Barbara Alice Mann, editor*

Iroquois on Fire: A Voice from the Mohawk Nation
*Douglas M. George-Kanentiio*

Native America, Discovered and Conquered: Thomas Jefferson, Lewis & Clark, and Manifest Destiny
*Robert J. Miller*

The Praeger Handbook on Contemporary Issues in Native America
Volume 1: Linguistic, Ethnic, and Economic Revival
Volume 2: Legal, Cultural, and Environmental Revival
*Bruce E. Johansen*

The Tainted Gift: The Disease Method of Frontier Expansion
*Barbara Alice Mann*

Frontier Newspapers and the Coverage of the Plains Indian Wars
*Hugh J. Reilly*

Lethal Encounters: Englishmen and Indians in Colonial Virginia
*Alfred A. Cave*

Reservation "Capitalism": Economic Development in Indian Country
*Robert J. Miller*

Land and Spirit in Native America
*Joy Porter*

# *Resource Exploitation in Native North America*

## A Plague upon the Peoples

**Bruce E. Johansen**

NATIVE AMERICA: YESTERDAY AND TODAY

PRAEGER™

An Imprint of ABC-CLIO, LLC
Santa Barbara, California • Denver, Colorado

**Library of Congress Cataloging-in-Publication Data**

Names: Johansen, Bruce E. (Bruce Elliott), 1950- author.
Title: Resource exploitation in Native North America : a plague upon the peoples / Bruce E. Johansen.
Description: Santa Barbara, California : Praeger, 2016. | Series: Native America: yesterday and today | Includes bibliographical references and index.
Identifiers: LCCN 2015031278 | ISBN 9781440831843 (hard copy : alk. paper) | ISBN 9781440831850 (ebook)
Subjects: LCSH: Indians of North America—Land tenure. | Indians of North America—Government relations. | Environmental policy—North America. | North America—Environmental conditions.
Classification: LCC E98.L3 J64 2016 | DDC 323.1197—dc23
LC record available at http://lccn.loc.gov/2015031278

ISBN: 978–1–4408–3184–3
EISBN: 978–1–4408–3185–0

20 19 18 17 16 1 2 3 4 5

This book is also available on the World Wide Web as an eBook.
Visit www.abc-clio.com for details.

Praeger
An Imprint of ABC-CLIO, LLC

ABC-CLIO, LLC
130 Cremona Drive, P.O. Box 1911
Santa Barbara, California 93116-1911

This book is printed on acid-free paper ∞

Manufactured in the United States of America

# *Contents*

# *Series Foreword*

As we prepare *Resource Exploitation in Native North America: A Plague upon the Peoples* for publication in 2016, the series "Native America: Yesterday and Today" has broached its second decade. Initiated to examine enduring issues in Native American studies in historical and contemporary context, this series began with Bruce E. Johansen's general history, *The Native Nations of North America* (2 vols., 2005). It has included such titles as Barbara Alice Mann's *George Washington's War on Native America* (2005), *Daughters of Mother Earth: The Wisdom of Native American Women* (2006), and *The Tainted Gift: The Disease Method of Frontier Expansion* (2009). Alfred A. Cave's *Lethal Encounters: Englishmen and Indians in Colonial Virginia* (2011) examined the cruel nature of the earliest British Virginia immigrants' treatment of Native peoples.

Robert J. Miller has explored contemporary Native American economic development in *Reservation "Capitalism": Economic Development in Indian Country* (2012) as well as the consequences of the Doctrine of Discovery in *Native America, Discovered and Conquered: Thomas Jefferson, Lewis & Clark, and Manifest Destiny* (2006). Hugh J. Reilly's *Frontier Newspapers and the Coverage of the Plains Indian Wars* (2010) described the lives of Native American peoples during nineteenth-century conflicts through the lens of contemporary newspaper reports. Douglas M. George-Kanentiio's *Iroquois on Fire: A Voice from the Mohawk Nation* (2006) provided an analysis of Akwesasne Mohawk life in recent times from one of its major journalistic and activist figures. Joy Porter explored the theoretical context of environmental themes in *Land and Spirit in Native America* (2012).

As as an indication of our series' quality, many of our titles have been published in paperback by the University of Nebraska Press. This is unusual; university presses do not often reprint trade press works. As Native Americans, many of our authors bring fresh perspectives to their writing from their lives along with their perspectives tested by daily reservation reality. Our work shares an honest, hard-edged quality.

Our new title describes life and death in Indian Country from environmental causes, a theme I have been following since my first book (*Wasi'chu: The Continuing Indian Wars*, 1979). When I began research on that book 40 years ago during a well-publicized "energy crisis," talk was being heard that Native Americans' lands would become "national sacrifice areas" for resource exploitation.

Today, with one-third of Environmental Protection Agency Superfund sites in the United States on Indian reservations, "national sacrifice" is no longer a rumor. From the (now-closed) uranium mines of Navajo Country to the PCB-laced turtles on the Akwesasne Mohawks' homeland in upstate New York, Native American peoples are resisting energy and resource colonization to survive. After several hundred Navajo uranium miners died, mining of "yellow dirt" is no longer legal there. Cleanup is underway at Akwesasne in a disjointed, halting manner.

In many other places, however, contamination continues. Canada may be worse than the United States, with its emphasis on such "uncoventional" oil sources as tar sands in Alberta, mining of which affects many Native peoples. The Arctic, an environment that seems so pristine to the unknowing, naked eye, is laced with so many persistent organic pollutants that mothers breast-feed their babies with caution, and eating fish can be toxic. Across the Arctic, a world based on ice is melting away as temperatures rise. Across North America, a struggle is underway involving Native peoples to protect a habitable environment, a vitally important issue for individual and community survival.

*—Bruce E. Johansen*
*Series Editor*
*Omaha, NE*
*July 2015*

# Introduction

Every book has a generative moment, but few are as easily fixed in time as this one. It was born with an e-mail of July 22, 2013, from Lisa Jones, a Navajo woman, who had read an Internet essay of mine, "The High Cost of Uranium in Navajo Country." What follows is an excerpt of a much longer, heartrending message in which she described how her extended family was being ravaged by radiation-induced cancers:

> Just want to start off with a thank you for the acknowledgement of the Navajo nation and all the data you capture. Unfortunately, I am witnessing it all first-hand. My name is Lisa Jones, my father and family comes from Wide Ruin, AZ and my mother's side is from Chambers/Sanders, AZ. I am sitting here at work in disbelief reading your article and other articles related to the uranium mining, as I will be leaving work soon to visit my uncle in the hospital as he battles his last days with cancer. . . . My uncle Jim Baldwin is 80 years old and has been in remission and back again. He is from the ranches in the Chambers area who has lived there [a] majority of his life with six kids, who now has an abundance of grandchildren and great grandchildren. My cousin Pauline Baldwin (Jim's daughter) had a tumor and is currently in remission. My aunt is currently battling breast cancer with life expectancy of four years, currently going through chemo and radiation with no insurance. My 44-year-old cousin Tanya Baldwin (Jim's niece) just passed from cancer Jan. 2013. My aunt Dorothy Baldwin Lugar has breast cancer and [is] still going through chemo. My aunt Phyllis Baldwin Montgomery McMahon is in remission from breast cancer. My grandmother Rose Baldwin just passed away May 2012. She also had stomach cancer, but lived a long life at 100. My grandmother's sister had

> stomach cancer. This is all on my mother's side . . . on the Navajo reservation in the Chambers and Sanders area . . . all my maternal family [has been] wiped out by cancer.

Lisa Jones's message completed a circle. I was working on my first book, *Wasi'chu: The Continuing Indian Wars*, in 1976, when Esther Keeswood of the Coalition for Navajo Liberation introduced me to both uranium and coal mining in Navajo Country. Women, especially elders, have been prominent in modern-day Native social and political movements. Emma Yazzie, who was more than 70 years of age when I met her, became an important Navajo leader against coal strip mining as her herds of sheep were devastated by mining pollution (Johansen and Maestas 1979, 143–44).

Yazzie led us to the floor of a strip mine near her hogan, disregarding the orders of mine security, to watch draglines at work, saying, "Don't go alone. They'll turn you away. They're afraid of me!" (Johansen and Maestas 1979, 145). Yazzie showed her disgust at the mining by unearthing surveyors' sticks and dumping them on supervisors' desks.

Yazzie found herself between coal strip mines and the Four Corners coal-fired electrical plant, which puts out a polluting plume so large that NASA astronauts observed it from Earth orbit. Her sheep had, by 1976, become small, skinny, and sickly, with gray wool. The Four Corners plant exports power to Los Angeles, Phoenix, Las Vegas, and other cities in the region. Yazzie's hogan had no electricity. The transmission lines buzzed above Emma Yazzie's hogan, which contained not so much as a single lightbulb.

When she gave me this tour, Yazzie was standing under the largest methane "hot spot" in North America. This area on and near the Navajo Nation, which stands out on satellite photographs taken between 2003 and 2009, has been created by coal mining and coal-fired power generation from the Four Corners Power Plant on the reservation and the San Juan Power Plant nearby. Some of this power generation has since been shut down to avoid expensive pollution controls. The methane hot spot, which covers about 2,500 square miles (roughly half the area of Connecticut) not far from the intersecting borders of Colorado, Arizona, New Mexico, and Utah, was described in the October 16, 2014, issue of *Geophysical Research Letters* (Kort et al. 2014). The Four Corners methane plume by itself makes up 10 percent of U.S. methane emissions, according U.S. Environmental Protection Agency (EPA) estimates (Minard 2014).

Yazzie did not know the academic language of resource colonization and exploitation, but she did know that the plume of smoke and ash that blew upslope near her hogan stunted the growth of her lambs and their dirty-brown coats. She called the mine and the power plant a monster and a harbinger of death.

During the 1940s, representatives of the companies that mined some of the uranium that powered the growing nuclear arsenal of the United States (and, later, early nuclear power plants) arrived on the Navajo Nation offering prosperity from mining jobs. Left unstated was the eventual cost of this uranium boom: an epidemic of cancer (especially lung cancer) in a place that had never known it. Until recently, cancer had no name in the Navajo language. Today, it rides the winds across the reservation as dust from huge piles of tailings (mining waste). The mining and milling of uranium is now illegal on Navajo land, but a deadly legacy remains.

Uranium mining has been only one part of energy colonization on the Navajo Nation that led the U.S. Department of Energy during the 1970s to call it a "national sacrifice area" (Johansen and Maestas 1979, 143, 170). A large part of this sacrifice involved the use of plentiful coal reserves under the Navajo Nation to generate electricity, thrumming along transmission lines to Los Angeles and other cities, over hogans that have no power, polluting the air that Emma Yazzie and many other Navajos breathe. Navajos continue to battle for a cleaner environment; in 2015, the EPA announced a $168 million settlement that may remedy part of the pollution over several years around the Four Corners plant that ruined Yazzie's flock 40 years ago. Neighbors of the plant are still suffering asthma and other health maladies with inadequate access to health care.

Near the town of Page, the Navajo Generating Station, the largest coal-fired plant in the American West, still pours plumes of carbon dioxide, lead, mercury, nitrogen oxide, and other metals out of three 775-foot stacks. Abraham Lustgarten of ProPublica reported in 2015 that "every hour the Navajo's generators spin, the plant spews more climate-warming gases into the atmosphere than almost any other facility in the United States. Alone, it accounts for 29 percent of Arizona's emissions from energy generation. The Navajo station's infernos gobble 15 tons of coal each minute, 24 hours each day, every day." According to the Navajo government, cancer rates in the area have doubled since the plant began operations four decades ago. "You are trying to raise your family in this environment, and you realize this is one of the top 10 dirtiest plants in the nation and it's been spewing all this stuff for 40 years," said Nicole Horseherder, a Navajo environmental activist. "Who is going to speak up and say, 'Look, we are paying a huge cost so that the state of Arizona can have its profits, have its taxes, have its electricity, have its water?'" (Lustgarten 2015).

When reservations (or, in Canada, reserves) were assigned to American Indians in the mid- to late nineteenth century, the fossil-fuel age was just beginning. The uses of uranium were largely unknown, and the federal governments of the United States and Canada, as well as corporations, had only the slightest inkling of what mineral and fuel riches lay under the reservations' often-barren soil. By the late twentieth century, however, these riches were being exploited, and many Native nations became virtual resource colonies.

Shortly before I met Emma Yazzie, a report from the Federal Trade Commission distributed during October 1975 said that an estimated 16 percent of the uranium reserves in the United States that were recoverable at market prices were on reservation lands; this was about two-thirds of the uranium on land under the legal jurisdiction of the U.S. federal government. There were, at that time, almost 400 uranium leases on these lands, according to the Federal Trade Commission, and between 1 million and 2 million tons of uranium ore a year, about 20 percent of the national total, was being mined from reservation land.

Resource colonization has continued. Coal, oil, coal, gas, and other minerals worth $3.5 billion were extracted from Native American lands in the United States during fiscal year 2011, compared to $2.8 billion worth in fiscal year 2010, according to a report by the Government Accountability Office (2012). Reservation governments received some monetary benefit from this activity: $538 million in fiscal year 2011 due to the increased activity, compared to $408 million in 2010. The report did not include the value of copper, gold, and other hard-rock resources from federal or Indian lands; mining companies are not legally obligated to pay royalties. They pay fees for leases. Resource exploitation of First Nations in Canada also has developed along similar lines. In 2014, American Indian revenues from mineral leases in the United States (34,000 owners in 34 tribes and nations) passed $1 billion for the first time, up about $200 million from 2013 ("Energy Revenues" 2014).

## *A PLAGUE UPON NATIVE PEOPLES*

This book examines environmental problems stemming from resource exploitation on Native lands from Nunavut and the Canadian Arctic southward through the United States into Mexico. It is mainly contemporary, but all of these problems have historical roots. The work focuses on Native resistance to resource colonization that is addressing the many environmental crises across Native North America that have become a plague on many indigenous peoples.

Pollution nearly inevitably follows resource exploitation. Native peoples in the United States today often live on ruined, exhausted land, suffering toxic consequences. Fully one-third of the Superfund sites declared by the EPA in the United States are on Native lands. This book examines a wide variety of serious environmental issues affecting scores of Native peoples in Alaska, Canada, and the United States.

Many reservation residents are dealing with illnesses due to the use of their lands for several decades as industrial dumps and mine sites. This work examines acute effects of exposure to dioxins, PCBs, and other persistent organic pollutants in several Native communities, most acutely in the Arctic, where Native consumption of sea life (their traditional diet) has been curtailed and Inuit mothers are sometimes warned not to breast-feed their infants. Mothers' milk may be toxic. This book also examines present and prospective effects of global warming (again, most acute in Arctic areas of Canada and Alaska) on Native peoples. Salmon, for example, are imperiled by warming water. The Arctic, which looks so pristine to the untutored eye, is also experiencing the world's most rapid rate of climate change. An Inuit culture based on ice is melting, and many Inuit hunters have been injured or killed by falling through thin ice.

Environmental provocations afflicting Native American peoples in the United States (a range of problems equal to those of any Third World nation) range from uranium mining to kitty litter for the Navajos to the devastation wrought by dioxin, PCBs, and other pollutants on the agricultural economy of the Akwesasne Mohawk reservation in northernmost New York State. As with the Akwesasne Mohawks, some of the most serious problems span international borders. The Yaquis, whose homelands span the U.S.–Mexican border, have been afflicted with some of the same pesticides as the Mohawks on the U.S.–Canadian border.

Some of the environmental problems faced by indigenous peoples in the United States strain one's sense of credulity. Witness the Eskimos of Point Hope, Alaska, who have learned that their land had once been proposed as the site of a new harbor to be created with nuclear weapons. The harbor was never created, but the Point Hope Eskimos still found themselves hosting uninvited nuclear waste. Other Alaskan Eskimos have found their reindeer rendered inedible, polluted with a number of heavy metals. The Western Shoshone of Nevada have come to call themselves "the most bombed nation on Earth," a reference to a neighboring test range for nuclear weapons and (more recently) a now stalled proposal to open a national waste-uranium repository at Yucca Mountain.

Canada, which often prides itself on an allegedly humane civil rights record regarding indigenous peoples, has become a major source of indigenous environmental contamination and conflict. The Innu of Labrador have been afflicted with sulfide mining, aluminum smelting, and noise pollution from squadrons of military aircraft. Some of the most intense resource exploitation in Canada takes place in remote locations, such as among the Lubicon Cree of northern Alberta, whose lands were so inaccessible in 1900 that treaty makers completely missed them. Today, roads have opened their lands to massive oil drilling and logging. The lands of the Cree in Quebec have been scarred by widespread dam building near James Bay that has contaminated large areas with toxic methylmercury as well as other pollutants. Uranium mining has decimated the Dene in Canada's Northwest Territories much as it has ravaged the Navajo in the southwestern United States. The catalog of indigenous environmental issues in Canada spans the range of resources—from hydropower to diamonds, uranium, gold, silver, sulfide, aluminum, oil, and natural gas. Several First Nations in Alberta have been scarred by the development of oil from tar sands.

Mexico's indigenous peoples have been dealing with a wide variety of environmental issues, from the Mayas of Chiapas in the South (who have been protesting the intrusion of roads, dams, logging, and hydropower dams under the Zapitista banner) to the Huicholes Indians of Sinaloa, Baja California, who harvest herbicide-laced tobacco. The workers and their families not only experience toxicity on the job; they also live in the fields, so their bodies are poisoned around the clock. Elsewhere in Mexico, the world's largest producer of silver has been ordered to cut production because children near its facilities were found to have unhealthy levels of lead in their bodies. Near Mexico City, in the state of Morelos, at the birthplace of the Mexica (Aztec) god Quetzalcoatl as well as Emilio Zapata, local indigenous peoples took to the streets to resist construction of a golf course that would have taken much of their water in a historically dry region. The image of Zapata was invoked by the protesters, who eventually defeated the proposed golf course.

## *AN ENVIRONMENTAL ETHOS*

In addition to being rooted in their homelands, Native Americans maintain historical, spiritual bonds to the land that foster attention to environmental threats. Longtime fishing rights activist Billy Frank Jr. said that this connection places protection of environment and its role in sustaining human and all other life "at the top of our priority list" (Russo 2013, 235). "When we say the Okanagan [native] word for 'ourselves,'" said Jeanette Armstrong, "we are actually saying the ones [that] are 'dream' and 'land' together" (Grossman and Parker 2012, 38). Chief Willie Charlie has said, "Mother Earth is crying, and we need to pay attention to what she is saying" (Grossman and Parker 2012, 45–46).

Activist and author Kurt Russo, who works with the Native American Land Conservancy, commented that "our courage to acknowledge this crisis, our conviction to stand up for unborn generations, our connection to nature and, through nature, to each other, and our resilience—as a family, as a species, as peoples—will determine whether we hear her cry for the agony of extinction or stand idly by and bear witness to a great dying" (Russo 2013, 236).

"Indigenous communities are, in general, in a unique position given their history and knowledge to understand and respond to the crisis," commented Russo. "[They] are, in general, more informed and engaged than the majority of Americans or their political and corporate leaders. . . . As place-based communities of inter-related families with historical consciousness, indigenous peoples are also more resilient, and thus able to face, rather than deflect or deny, the true magnitude of the crisis [that] will have long-lasting and potentially catastrophic consequences for every life-form and human community" (Russo 2013, 234).

On September 29, 2011, Daniel Wildcat, Yuchi member of the Muscogee Nation of Oklahoma and a professor at Haskell Indian Nations University in Lawrence, Kansas, speaking at the Center of the American West of the University of Colorado, Boulder, said that, until recently, Native people were members of tribes, not nation-states, with a relationship to nature that defined species on which people depended for survival in familial terms as relatives, not as exploitable resources. The bison on the Great Plains, salmon among the Coast Salish, and corn across much of today's North America (Turtle Island), was "the central relative we acknowledged" (Berry 2011).

Indigenous environmental activism (including visceral opposition to development of extractive industry) stems from long historical experience with resource colonization, which works in synthesis with a spiritual ethos that invests animus in everything natural. While European religions often restrict their blessings to humanity, many Native Americans interpret "all my relations" to mean all of nature—animate and not—even the rocks on which we walk. This respect for nature is fundamental and enduring and at the root of traditional Native American responses to economic development. Definitions of "balance" are couched in this context, counterpoising protection of "mother earth" with an immigrants' ethos that seeks a "mother lode," rooted in *Genesis* ("Go forward, multiply, and subdue the Earth"). Naomi Klein, in *This Changes Everything:*

*Capitalism and the Climate* (2014, 183) writes of industrialists who "view . . . nature as a bottomless vending machine."

Awareness of economic development's costs also animates the non-Indian environmental movement. "This, then, is the nemesis that modern Western man, together with his imitators . . . has brought upon himself by following the directive given in the first book of *Genesis*," wrote the great English historian Arnold Toynbee. "That directive has turned out to be bad advice, and we are beginning, wisely, to recoil from it" (Toynbee 1973, n.p.).

## *RECOGNITION AND REMEDY*

The non-Native world is slowly recognizing the toll that its appetites for natural resources wreak on indigenous peoples. In February 2013, UN Special Rapporteur on the Rights of Indigenous Peoples James Anaya issued a request "for information on extractive and energy industries in or near indigenous territories . . . and the human rights issues that these industries generate . . . such as mining, petroleum or gas projects and energy production, including hydro-electric projects." A report to be compiled by Anaya's office "will identify good practices for the industries to use in order to avoid or overcome . . . human rights violations." This report was compiled because, according to the Indian Country Today Media Network, "sovereign Indian nations, particularly in the American West, continue to be significantly impacted by radioactive and other hazardous wastes because of the proximity of nuclear test sites, uranium mines, power plants, toxic waste dumps and extractive industries" (Toensing 2013).

In addition to corporations, the U.S. military has been a source of pollution in Indian Country, according to Gregory Hooks and Chad L. Smith, who coauthored "The Treadmill of Destruction: National Sacrifice Areas and Native Americans" in the *American Sociological Review*. The study found "that much of the disproportional exposure of Native Americans to environmental dangers throughout the 20th century was the result of militarism, rather than economic competition. And it shows that historically coercive governmental policies in locating Indian reservations are a major factor in determining their exposure" (Hooks and Smith 2004, 558).

## *OUTLINE OF THE WORK*

Chapter 1, "Native America's Radioactive Legacy," notes that uranium mined from Native American lands supplied a substantial proportion of the fuel for early nuclear power plants as well as the U.S. nuclear arsenal. By the 1970s, many of the early miners were dying of lung cancer, most notably among the Navajo and the Dene of the Canadian Northwest Territories. In Washington State, nuclear waste from the Hanford plants afflicted the Yakamas. Uranium mining also has caused a plague of cancer among the Laguna Pueblo. The Navajos organized to stop uranium mining and milling, and today both are illegal there—but only after several hundred people had died.

Chapter 2, "Toxic Chemicals: A Plague upon the People," begins in Nunavut, the semisovereign Inuit nation in northern Canada where it is snowing poison, as PCBs, dioxins, and other chemicals produced by southern industries are swept northward by prevailing winds. Southward, at Akwesasne, in northernmost New York State and nearby Canada, the fish, laced with the same chemicals, are often inedible. The Huicholes of Mexico live in a land laced with them. In all of these locations, Native peoples are organizing to restore a livable environment. The Inuit, in particular, have taken an active role in efforts to outlaw use of these toxins in an international agreement: the Stockholm Convention.

The imaginations of dam builders have only rarely been restrained by the fact that land that they want to flood is owned and occupied by Native Americans. Chapter 3, "Canada's Hydroelectric Dreams," begins in the 1970s, when Hydro-Quebec, an unwelcomed guest, came to stay on Cree land in northern Quebec, proposing a vast network of dams that began to transform the Cree homeland, introducing toxic methylmercury. The Crees organized, enlisted support worldwide, and fought off the threat. In the meantime, the developers were making plans in Manitoba.

Global warming affects everyone, some more rapidly and intensely than others. Chapter 4, "Climate Change and Native Peoples," surveys the Canadian and Alaskan Arctic, which are among the most quickly warming areas on Earth. Alaskan Native communities face climate-induced change, including the relocation of entire coastal villages. Elsewhere in Native America, elders tap generational memory to tell scientists how their lives have been reshaped by a rapidly warming climate.

Chapter 5, "Welcome to My Moonscape," discusses Native peoples in Alberta who have found some of their lands devastated by tar (or oil) sands mining. Native peoples in the United States have taken a leading role in opposing the Keystone XL Pipeline, which is being proposed to carry tar sands oil from Alberta to the U.S. Gulf Coast. Blockades have been set up by the Nez Perce (in Idaho) and Lakota (at Pine Ridge, South Dakota) to block construction materials on their way to Alberta, provoking many arrests. Pine Ridge draws its water from the planned path of the pipeline. On one occasion in Idaho in 2013, the tribal chairman and most of the Nez Perce council were arrested (along with two dozen other people) as police dispersed a crowd of more than 200. Others, including Mik'maq bands in New Brunswick, have put their bodies on the line, with arrests, to impede fracking (hydraulic fracturing), in which high-pressure chemicals are injected deep underground to pulverize shale so that oil may be pumped out.

Chapter 6, "Mining: Mother Earth or Mother Lode?," describes several of many instances in which Native peoples across North America have organized against mining on their homelands, notably in Alaska, where large salmon runs could have been imperiled by the proposed Pebble Mine before a coalition of Native peoples and environmentalists obtained governmental restraint. In Wisconsin, copper, zinc, and sulfide mining are being proposed and resisted by Native peoples. Silver mining in Mexico is poisoning indigenous children. Coal strip mining is an important issue for Hopis and Navajos, as some shepherds have lost many of their sheep to the toxic plumes of the Four Corners power plant.

This book links present-day Native American cultural and economic revival to a fundamental struggle to restore the health of both Native peoples and their homelands. It links past and present with a sense of Native Americans' perceptions of nature and the sacred and, in so doing, also provides the majority society with an example to emulate as we emerge, by necessity, from the age of fossil fuels into a sustainable energy paradigm.

## *FURTHER READING*

Berry, Carol. 2011. "Scholar Daniel Wildcat in Discussing the Environment: It's Relatives, Not 'Resources.'" Indian Country Today Media Network, October 1. http://indiancountrytodaymedianetwork.com/gallery/photo/scholar-daniel-wildcat-in-discussing-the-environment%3A-it%E2%80%99s-relatives%2C-not-%E2%80%98resources%E2%80%99-56599

Brugge, Doug, Timothy Benally, and Esther Yazzie-Lewis, eds. *The Navajo People and Uranium Mining*. Albuquerque: University of New Mexico Press, 2006.

Eichstaedt, Peter. *If You Poison Us: Uranium and American Indians*. Santa Fe, NM: Red Crane Books, 1994.

"Energy Revenues to American Indians Top $1 Billion for First Time." 2014. Indian Country Today Media Network, December 16. http://indiancountrytodaymedianetwork.com/2014/12/16/energy-revenues-american-indians-top-1-billion-first-time-158312

Gedicks, Al. *The New Resource Wars: Native and Environmental Struggles against Multinational Corporations*. Boston: South End Press, 1993.

Government Accountability Office. 2012. "GAO Report: $3.5 Billion Worth of Minerals Produced on Indian Lands in FY 2011." Indian Country Today Media Network, December 13. http://indiancountrytodaymedianetwork.com/article/gao-report-35-billion-worth-minerals-produced-indian-lands-fy-2011-146271

Grinde, Donald A., Jr., and Bruce E. Johansen. *Ecocide of Native America: Environmental Destruction of Indian Lands and Peoples*. Santa Fe, NM: Clear Light, 1995.

Grossman, Zoltán, and Alan Parker. 2012. *Asserting Native Resilience: Pacific Rim Indigenous Nations Face the Climate Crisis*. Corvallis: Oregon State University Press.

Hooks, Gregory, and Chad L. Smith. 2004. "The Treadmill of Destruction: National Sacrifice Areas and Native Americans." *American Sociological Review* 69, no. 4: 558–75.

Johansen, Bruce E., and Roberto Maestas. 1979. *Wasi'chu: The Continuing Indian Wars*. New York: Monthly Review Press.

Klein, Naomi. 2014. *This Changes Everything: Capitalism and the Climate*. New York: Simon & Schuster.

Kort, Eric A., Christian Frankenberg, Keeley R. Costigan, Rodica Lindenmaier, Manvendra K. Dubey, and Debra Wunch. 2014. "Four Corners: The Largest U.S. Methane Anomaly Viewed from Space." *Geophysical Research Letters* 41, no. 19 (October 16): 6898–6903.

Lustgarten, Abraham. 2015. "Inside the Power Plant Fueling America's Drought." ProPublica in Indian Country Today Media Network, July 3. http://indiancountrytodaymedianetwork.com/2015/07/03/inside-power-plant-fueling-americas-drought-160928

Minard, Anne. 2014. "Methane 'Hot Spot' Seen from Space Hovers over Four Corners." Indian Country Today Media Network, October 28. http://indiancountrytodaymedianetwork.com/2014/10/28/methane-hot-spot-seen-space-hovers-over-four-corners-157560

Russo, Kurt. 2012. "Review of *Asserting Native Resilience: Pacific Rim Indigenous Nations Face the Climate Crisis*." *American Indian Culture and Research Journal* 37, no. 2: 234.

Toensing, Gale Courey. 2013. "Anaya Seeks Information for Study on Extractive and Energy Industries." Indian Country Today Media Network, February 6. http://indiancountrytodaymedianetwork.com/2013/02/06/anaya-seeks-information-study-extractive-and-energy-industries-147495

Toynbee, Arnold. 1973. "The Genesis of Pollution," *New York Times*, September 16, n.p.

# Chapter 1

## *Native America's Radioactive Legacy*

From the Dine (whom the Spanish called Navajo) to their namesake Dene of Canada's Northwest Territories (and many points between), Native peoples were recruited beginning in the early 1950s to mine "yellow dust"—uranium—and then, over decades, died in large numbers of torturous cancers. Uranium-induced cancers have become the deadliest plague unleashed on Native peoples of North America. On Navajo land, its mining and milling are now illegal, as Native people heed their "original instructions": some resources are better left in the ground.

Uranium mined from Native American lands supplied a substantial proportion of the fuel for early nuclear power plants as well as the U.S. nuclear arsenal. By the 1970s, many of the early miners were dying of lung cancer. In Washington State, nuclear waste from the Hanford plants afflicted the Yakamas. Uranium mining also has caused a plague of cancer among the Laguna Pueblo. The Navajos succeeded in stopping uranium mining and milling only after several hundred people had died of its effects and many more had suffered the tortures of cancers that once were nearly unknown in their country.

By 2014, at least 350 to 400 former Navajo underground uranium miners had died from maladies caused in large part by exposure to radiation, according to Chris Shuey, an environmental health researcher with the Southwest Research and Information Center (Knight 2013). Many more Native people also had died of a wide variety of malignant cancers, not usually from mining uranium themselves but because they lived with the "yellow dirt" in windswept waste (tailings) piles that have blown into every crack and crevice, indoors and outdoors, for decades.

In addition to the mining and milling of nuclear fuel, New Mexico also supplied for the United States its first test site for the atomic bomb, detonated on July 16, 1945, on a 100-foot tower. The Trinity Site on today's U.S. Army's

White Sands Missile Range is surrounded by two Apache tribes, several Navajo communities, and 19 American Indian pueblos. Reports at the time by American Indians and government witnesses described a light ashfall for days after the explosion, the effects of which were not investigated until 2014, when the National Cancer Institute started a study of radiation levels in New Mexico from that first test blast. With no prior evidence of effects, no one has been eligible for compensation under the Radiation Exposure Compensation Act, which covers other nuclear and uranium workers as well as "down-winders" who were exposed to radiation from later atomic tests (Lee 2014).

### *URANIUM'S HUMAN PRICE IN NAVAJO COUNTRY*

In 1950, Paddy Martinez, a Navajo sheepherder, brought a strange-looking yellow rock into Grants, New Mexico, from nearby Haystack Butte, initiating a mining boom. Navajo uranium miners at first hauled radioactive uranium ore out of the earth as if it were coal or any other mineral. Some ate their lunches in the mines and slaked their thirst with radioactive water. Their families' homes sometimes were built of radioactive earth, and their neighbors' sheep may have watered in small ponds that formed at the mouths of abandoned uranium mines. On dry, windy days, the gritty dust from uranium waste-tailings piles covered everything in sight.

About half the recoverable uranium in the United States lies in New Mexico—and about half of that is beneath the Navajo Nation. The Navajo language has no word for "radioactivity." Initially, no one told the miners that within two or three decades, many of them would die of radiation-induced cancers. In their rush to profit from uranium mining (and the lives of the miners), very few mining companies provided ventilation in the early years. Some miners worked as many as 20 hours a day, entering the mines just after the blasting of sandstone had filled the mines with silica dust. Many mine owners didn't even provide toilet paper. When miners relieved themselves, they wiped with fists of radioactive "yellowcake"—uranium ore.

The Kerr-McGee Company, the first corporation to mine uranium on Navajo Nation lands (beginning in 1948), found the reservation location extremely lucrative. There were no taxes at the time; no health, safety, or pollution regulations; and few other jobs for the many Navajos who had recently arrived home from service in World War II. Labor was cheap, and uranium, in demand for stockpiles of nuclear armaments, was expensive.

The first uranium miners in the area, almost all of them Navajos, remember being sent into shallow tunnels within minutes after blasting. They loaded the radioactive ore into wheelbarrows and emerged from the mines spitting black mucus from the dust, coughing so hard that many of them experienced headaches. Such mining practices exposed the Navajos who worked for Kerr-McGee to between 100 and 1,000 times the level of radon gas later considered safe. Officials for the U.S. Public Health Service (PHS) estimated these levels of exposure after the fact; in the earliest days, no one was monitoring the Navajo miners' health.

Carrie Arnold (2014) wrote in *Environmental Health Perspectives*, published by the National Institutes of Health, that the miners and their families "were not told that the men who worked in the mines were breathing carcinogenic radon gas and showering in radioactive water, nor that the women washing their husbands' work clothes could spread radionuclides to the rest of the family's laundry" after they had worked in the 521 now-abandoned uranium mines on the reservation ranging in size from "dog holes" that could accommodate only a single man to large mines from which radioactive ore was extracted in carts on rails.

Arnold wrote that health workers were allowed to interview uranium miners only after they "had to strike a Faustian bargain with the mining companies: They could not inform the miners of the potential health hazards of their work. Seeing it as the only way to convince government regulators to improve safety in the mines, the researchers accepted. The PHS monitored the health of more than 4,000 miners between 1954 and 1960 without telling them of the threat to their health. By 1965, the investigators reported an association between cumulative exposure to uranium and lung cancer among white miners and had definitively identified the cause as radiation exposure." The effects had been no secret, even early as 1950, when government workers monitored radiation levels in the mines that were as much as 750 times the limits deemed acceptable at that time.

## *"RADIOACTIVE FROM THE INSIDE OUT"*

More than 99 percent of the rock that the mines produced was waste, cast aside as tailings near mine sites after the uranium had been extracted. One of the mesa-like waste piles grew to be a mile long and 70 feet high. On windy days, dust from the tailings blew into local communities, filling the air and settling into water supplies. At the time, beginning during the 1950s, the Atomic Energy Commission (AEC) assured worried local residents that the dust was harmless.

When mining was initiated, no one considered environmentally appropriate ways to deal with its tailings piles. Even if the tailings were to be buried—a staggering task—radioactive pollution could leak into the surrounding water table. A 1976 Environmental Protection Agency (EPA) report found radioactive contamination of drinking water on the Navajo Reservation in the Grants, New Mexico, area near a uranium mining and milling facility (Eichstaedt 1994, 208).

Arnold (2014) wrote of the mine tailings' legacy more than half a century after the first uranium was mined:

> In a low, windswept rise at the southeastern edge of the Navajo Nation, Jackie Bell-Jefferson prepares to move her family from their home for a temporary stay that could last up to seven years. A mound of uranium-laden waste the size of several football fields, covered with a thin veneer of gravel, dominates the view from her front door. After many years of living next to the contamination and a litany of health problems she believes it caused, Bell-Jefferson and several other

> local families will have to vacate their homes for a third round of cleanup efforts by the U.S. Environmental Protection Agency.

Members of Bell-Jefferson's family had used radioactive water for years, "splashing and swimming in pools of radioactive water that had been pumped out of the mines and then collected on their property. The contaminated water looked and tasted perfectly clean. Families used it for cooking, drinking, and cleaning. Hogans and corrals were built with mine wastes, as were roads" (Arnold 2014).

Gilbert Badoni of Shiprock, New Mexico, said that as a child, he played in the tailings with siblings and drank radioactive water during the decades his father worked in uranium mines in Colorado in the 1950s and 1960s. Badoni said that his father arrived at home after work covered in yellow uranium dust, which covered everything in their small home as their mother brushed it off his clothes. Badoni blamed his lung problems and his siblings' cancers on that exposure. "The U.S. government has abused innocent women and children. They have abused my family," Badoni said, choking back tears. "They have abused my Navajo people. That's not right" ("Uranium Miners" 2000).

Uranium mine dust produced silicosis in the miners' lungs in addition to lung cancer and other problems associated with exposure to radioactivity. By the 1960s, nearly 200 miners already had died of uranium-related causes. That number had doubled by 1990. Radioactivity also contaminated drinking water in parts of the Navajo reservation, producing birth defects and Down syndrome, both previously all but unknown among the Navajo. Of all infant deaths in areas served by the Navajo Indian Health Service area for the years 1990 through 1992, 35 percent were caused by congenital anomalies. Mortality attributed to malignant neoplasms (age-adjusted rate) was 78.5 percent in the years from 1990 through 1992 (Colomeda 1998). A 1976 EPA report found radioactive contamination of drinking water on the Navajo reservation in the Grants area near a uranium mining and milling facility.

The government knew of the risk at least by 1978, however, when the Department of Energy released a Nuclear Waste Management Task Force report disclosing that people living near the tailings piles ran at least twice the risk of lung cancer as the general population. Even then, the Coalition for Navajo Liberation was aware that a number of miners were dying of lung cancer. As Navajo miners continued to die, children who played in water that had flowed over or through abandoned mines and tailings piles came home with burning sores. Downwind of uranium processing mills, the dust from yellowcake sometimes was so thick that it stained the landscape half a mile away.

Dry winds blew dust from tailings piles through the streets of many Navajo communities. "We used to play in it," said Terry Yazzie of an enormous tailings pile behind his house. "We would dig holes and bury ourselves in it" (Eichstaedt 1994, 11). The neighbors of this particular tailings pile were not told it was dangerous until 1990, 22 years after the mill that produced the pile had closed and 12 years after Congress authorized the cleanup of uranium mill tailings in Navajo country. Abandoned mines also were used as shelter by animals that inhaled radon and drank contaminated water. Local people milked the animals and ate their contaminated meat.

Peter Eichstaedt wrote that miners watched members of their families die because of radiation poisoning that permeated their entire lives. Some miners were put to work packing thousand-pound barrels of yellowcake. Some of the miners ingested so much of the dust that it was "making the workers radioactive from the inside out" (Eichstaedt 1994, 11).

## *OPPOSITION TO URANIUM MINING*

Through opposition to uranium mining among Indians and non-Indians alike runs a deep concern for the long-term poisoning of land, air, and water by low-level radiation. By the 1970s, these concerns had provoked demands from Indian and white groups for a moratorium on all uranium mining, exploration, and milling until the issues of untreated radioactive tailings and other waste-disposal problems were faced and solved. Doris Bunting of Citizens Against Nuclear Threats (a predominantly non-Native group that joined with efforts with the Coalition for Navajo Liberation [CNL] and the National Indian Youth Council in opposition to uranium mining) supplied data indicating that radium-bearing sediments had spread into the Colorado River basin, from which water is drawn for much of the Southwest.

The specter of death that haunted former Navajo uranium miners came at what company public relations specialists might have deemed an inappropriate time. By late 1978, more than 700,000 acres of Indian land were under lease for uranium exploration and development in an area centering on Shiprock and Crownpoint, both on the Navajo Nation. Atlantic Richfield, Continental Oil, Exxon, Humble Oil, Homestake, Kerr-McGee, Mobil Oil, Pioneer Nuclear, and United Nuclear were among the companies exploring, planning to mine, or already extracting ore. During the 1980s, the mining frenzy subsided somewhat as recession and a slowing of the nuclear arms race reduced demand. Some ore was still being mined, but most of it lay in the ground, waiting for the next upward spike in the market.

Even as the boom ended, many of the miners who had been condemned to slow death by lung cancer did not yet know that the yellow ore they had mined was killing them. Peter Eichstaedt's painstakingly detailed study described how the miners learned what had been done to them and how they went about winning at least a possibility of compensation from the U.S. government. Using a combination of interviews and scientific data, Eichstaedt demonstrated that the U.S. government (particularly the AEC) knew that uranium mining was poisoning the Navajos almost from the beginning, a key point in the debate over compensation that has thus far resulted in a small proportion of miners collecting $100,000 each. The government and the mining companies kept medical knowledge from the miners themselves out of concern for national security and profits.

## "ORIGINAL INSTRUCTIONS," URANIUM MINING, AND NAVAJO CULTURAL VALUES

The mining of uranium became a major political and cultural issue among the Navajo during the 1970s as the cancer death toll mounted. Lung cancer became so widespread that it had major cultural implications, provoking many Navajo to reexamine traditional teachings about the need for harmony with regard to the Earth. Following debate centered on this examination, uranium mining and milling henceforth became illegal in the Navajo Nation.

Activists in the CNL sought to have uranium mining and milling outlawed on the reservation as soon as its human toll became evident, a position that eventually prevailed in the Navajo government after a quarter of a century of debate and industry lobbying. Esther Keeswood, a member of the CNL from Shiprock, a reservation town near several tailings piles, said in 1978 that the CNL had documented the deaths of at least 50 residents (including several uranium miners) from lung cancer and related diseases. The number of deaths increased steadily after that.

Uranium's yellow dirt is part of the natural order of things in Navajo Country, and a ban on its mining and milling was regarded by traditional people as a cultural victory, a return to the original instructions of creation. The yellow dirt in mythology is believed to be the antithesis of the life-giving yellow powder of corn pollen:

> In one of the stories the Navajos tell about their origin, the *Dineh* (the people) emerged from the third world into the fourth and present world and were given a choice. They were told to choose between two yellow powders. One was yellow dust from the rocks, and the other was corn pollen. The Dineh chose corn pollen, and the gods nodded in assent. They also issued a warning. Having chosen the corn pollen, the Navajo were to leave the yellow dust in the ground. If it was ever removed, it would bring evil. (Eichstaedt 1994, 47)

In this cultural interpretation, the commercial use of the yellow dirt turned it into *Leetso*, a powerful monster that inflicted punishment on the Navajo people in the form of disease and suffering. The only way to restore harmony with nature (a very important attribute in Navajo culture) is *hozho nashaadoo* ("to walk in harmony") (Csordas 1999). The only proper response was to restore harmony by banning the mining and milling of uranium. The yellow dirt is a poison that disrupts the gathering of sacred herbs in contaminated areas. The Navajo seek to balance elements of land, water, and sunlight (fire) (Woody et al. 1981). Some of the elders blamed themselves for allowing disruption of *hozho nashaadoo*.

## THE LARGEST URANIUM SPILL IN THE UNITED STATES

The largest expulsion of radioactive material in U.S. history occurred at 5:00 A.M. on July 16, 1979, in the Navajo Nation. On that morning, more than 1,100 tons of uranium mining wastes—tailings mixed with water—gushed through a breached packed-mud dam near Church Rock, New Mexico. With the tailings, 100 million gallons of radioactive water flowed through the dam before the crack was repaired.

By 8:00 A.M., radioactivity was monitored in Gallup, New Mexico, nearly 50 miles away. The contaminated Rio Puerco showed 7,000 times the allowable standard of radioactivity for drinking water below the broken dam shortly after the breach was repaired, according to the Nuclear Regulatory Commission (NRC). The few newspaper stories about the spill outside of the immediate area noted that the area was sparsely populated and that the spill posed no immediate health hazard. The New Mexico Environmental Improvement Division said that while the spill had been "potentially hazardous . . . its short-term and long-term impacts on people and the environment were quite limited" (Johansen 1997, 11). With these soothing words, the same reports also recommended that ranchers in the area avoid watering their livestock in the Rio Puerco. After the Rio Puerco spill, however, several Navajos said that calves and lambs were born without limbs as well as other severe birth defects. Other livestock developed sores, became ill, and died after drinking from the river.

The same report noted that the river water was not being used for human consumption and that "the extent to which radioactive and chemical constituents of these waters are incorporated in livestock tissue and passed on to humans is unknown and requires critical evaluation" (Johansen 1997, 12). The report also said that the accident's effect on groundwater should be studied more intensely. Tom Charley, a Navajo, told a public meeting at the Lupton Chapter House that "the old ladies are always to be seen running up and down both sides of the [Rio Puerco] wash, trying to keep the sheep out of it" (Johansen 1997, 12). The Centers for Disease Control examined a dozen dead animals and called for a more complete study in 1983, then dropped the subject.

Even as the Rio Puerco ran radioactive, the town of Grants styled itself as the "Uranium Capital of the World" as new pickup trucks appeared on the streets and mobile home parks grew around town, filling with non-Indian workers in the uranium boom. For several years, before the boom abruptly ended in the early 1980s, many workers in the uranium industry earned $60,000 or more a year (about $180,000 in 2014 dollars). Following a collapse of demand for uranium in the early 1980s, Grants dropped the nickname "Uranium Capital of the World" and began promoting itself as a haven for retirees under the new slogan "Grants Enchants."

More problems began to appear. A waste pile at the United Nuclear mill that had produced the wastes that gushed down the Rio Puerco in 1979 was detected leaking radioactive thorium into local groundwater. On May 23, 1983, the state of New Mexico issued a cease-and-desist order to United Nuclear to halt the radioactive leakage. The company refused to act, stating that its leak did not violate state regulations. Allendale and Appalachian, two insurance companies that were liable for roughly $35 million in payments to United Nuclear because of losses related to the accident, sued the company on the belief that it knew the that dam that burst was defective before the spill. The dam was only two years old at the time of the accident.

Along the Rio Puerco, several ranchers reacted to state assurances that the spill left no long-term effects by selling their land, for millions of dollars, to the federal government. The ranchers sold out under the 1974 Relocation Act, meant to move Navajos from the former "joint-use area" claimed by the Hopis. The land was purportedly acquired to relocate Navajos who had lost their homes in the land dispute with the Hopis. The Navajos asked the EPA, the Bureau of Indian Affairs, and the Relocation Commission for assurance that the land was safe. All three declined to provide the requested written assurance to the Navajos. The Navajos raised several questions, including the extent of contamination in underground aquifers, the extent of remaining radioactivity in surface waters and soils, the effects of wind-blown dust from the contaminated area, and the long-term effects of the contamination on livestock and people in the area.

Uranium contamination of river water was one problem. Lack of groundwater for wells was another side effect of uranium and other mining. As a result of mining generally, the U.S. Geological Survey predicted that the water table at Crownpoint would drop 1,000 feet and that it would return to present levels only 30 to 50 years after the mining ceased. Much of what water remained could be polluted by uranium residue, the report indicated. The Indians owned the surface rights; the mineral rights in the area are owned by private companies, such as the Santa Fe Railroad. "If the water supply is depleted, then this [Crownpoint] will become a ghost town," said Joe Gmusea, a Navajo attorney. "The only people left will be the ones who come to work in the mines" (Johansen 1997, 11).

John Redhouse, associate director of the Albuquerque-based National Indian Youth Council, said that the uranium boom is "an issue of spiritual and physical genocide. . . . We are not isolated in our struggle against uranium development. . . . Many Indian people are now supporting the [anti–uranium mining] struggles of the Australian aborigines and the Black indigenous peoples of Namibia [South West Africa] against similar uranium developments. We have recognized that we are facing the same international beast" (Johansen 1997, 11).

## *THE STRUGGLE FOR COMPENSATION*

By 1978, the Navajos were beginning to trace the roots of a lung cancer epidemic that had perplexed many of them, as the disease had been rare among them before World War II. Harris Charley, who worked in the mines for 15 years, told a U.S. Senate hearing in 1979, "We were treated like dogs. There was no ventilation in the mines." Pearl Nakai, daughter of a deceased miner, told the same hearing, "No one ever told us about the dangers of uranium" (Grinde and Johansen 1995, 214). The Senate hearings were convened by Senator Pete Domenici (R–NM), who was seeking compensation for disabled uranium miners and for the families of the deceased. "The miners who extracted uranium from the Colorado Plateau are paying the price today for the inadequate health and safety standards that were then in force," Domenici told the hearing, held at a Holiday Inn near Grants (Grinde and Johansen 1995, 214).

Senate hearings during 1979 initiated proposals to compensate the miners for what investigators called deliberate negligence. By 1930, radioactivity in uranium mines had been associated with lung cancer by tests in Europe. Scientific evidence linking radon gas to radioactive illness existed after 1949, but measures to ventilate the Navajo mines were never implemented, as the government pressured Kerr-McGee and other producers to increase the amount of uranium they were mining. The PHS recommended ventilation in 1952, but the AEC said that it bore no responsibility for the mines despite the fact that it bought more than 3 million pounds of uranium from them in 1954 alone.

Dr. Joseph Wagoner, special assistant for occupational carcinogens at the Occupational Safety and Health Administration, a federal agency, said that of 3,500 persons who mined uranium in New Mexico, about 200 had died of cancer by the late 1970s. In an average population of 3,500 persons, 40 such deaths could be expected. The 160 extra deaths were not the measure of ignorance, he said. Published data regarding the dangers of radon were widely available to scientists in the 1950s, according to Wagoner. Health and safety precautions in the mines were not cost effective for the companies, he said. "Thirty years from now we'll have the hidden legacy of the whole thing," Wagoner told Molly Ivins (1978, n.p.).

For a dozen years, bills that would have compensated the Navajo miners were introduced, discussed, and left to languish and die in congressional committees. Compensation was finally approved after more than a decade of congressional debate. Next, the federal bureaucracy stood in the way. By 1990, the cancer-related death toll among former miners had risen to 450 and was still rising. By the early 1990s, about 1,100 Navajo miners or members of their families had applied for compensation related to uranium exposure. The bureaucracy had approved 328 cases, denied 121, and withheld action on 663, an approval rate that Representative George Miller, chairman of the House Natural Resources Committee, characterized as "significantly lower than in other cases of radiation compensation" (Johansen 1997, 12). Representative Miller said that awards of compensation were being delayed by "a burdensome application system developed by the Department of Justice" (Johansen 1997, 12).

## *DELAYS IN COMPENSATION*

By the beginning of 2001, former Navajo uranium miners and millers, as well as their families, were holding meetings and protest marches in a renewed attempt to collect compensation under the Radiation Exposure Compensation Act for cancers associated with radiation exposure. Meetings and protest marches were held in Grants; in Cortez, Colorado; and in or near several towns in the Navajo Nation. National legislation compensating victims of uranium mining and fallout had been enacted in 1990, guaranteeing up to $100,000 to miners and down-winders who had suffered radiation exposure during nuclear weapons tests. The program was not funded until 1992 and even then only at very low levels. It ran out of money during 1993, when the government began issuing IOUs.

During July 2000, President Bill Clinton signed an amendment to the 1990 law extending its provisions to millers and transporters of radioactive uranium ore, once again without adequate funding to implement the law's provisions. In the meantime, increasing numbers of former uranium miners were suffering illnesses and deaths due to their exposure. The amendments expanded the list of eligible afflictions and reduced the amount of time a miner was required to have spent working with uranium to be eligible for the compensation program. The measure also opened the program to those who had worked in open-pit uranium mines and uranium milling plants as well as underground mines. Many miners had complained that they were excluded by the overly stringent rules of the original law.

Melton Martinez, who has been instrumental in the struggle to gain compensation for miners and millers, said that the government knew that uranium had caused many cancers and organ failures. Even wives of workers who washed the yellow-coated clothes of their miner husbands were experiencing cancers, Martinez said (Purdom 2001). Mayor Elisabeth Lopez-Rael of Milan, New Mexico, told a gathering there that she knew a man in Grants who was given an IOU by the government. "He told me he probably wouldn't live long enough to spend it and he has lost a lot of friends already," Lopez-Rael said (Purdom 2001).

Several former miners raised the possibility that aquifers in the area remained contaminated by radioactivity from the former uranium mines. One man asked, "Where does the contamination go? Into the water." Because of radioactivity's persistent nature, Martinez said, "we are doing this not only for ourselves and our neighbors, we are doing this for our children too" (Purdom 2001). "They did experiments on us like guinea pigs. It makes me angry," one former miner said as he sat on the steps outside the U.S. House of Representatives. "I would have lived longer, but they gave me a shorter life on this Earth" ("Uranium Miners" 2000).

## *CLEANUP PLANNED AT A FEW SPILL SITES AFTER THREE DECADES*

After almost three decades of appeals from the Navajo Nation, the EPA approved plans in September 2011 to clean up the Northeast Church Rock Mine near Gallup, the largest abandoned uranium mine in Navajo territory. The mine was used between 1967 and 1982 by the United Nuclear Corporation, which left behind roughly 1.4 million tons of uranium and radium-contaminated soil near its 125-acre uranium mill site. This mill and mine was the site of the largest release of radioactive waste in the United States. On July 16, 1979, a dam that had been restraining mounds of waterlogged tailings failed, flooding into the Rio Puerco in New Mexico downstream to Chambers, Arizona. According to the Navajo Nation, the spill, "combined with more than 20 years of discharges of untreated and poorly treated uranium mine water, has contributed to long-term contamination of the Puerco River in New Mexico and Arizona" ("Cleanup" 2011).

The area has long been designated a Superfund site by the EPA, comanaged by the NRC. The site is near the border of the Navajo Nation, on trust territory, but the mill site is privately owned. A small number of people live near the site, downwind and downstream of mine tailings waste piles. The people in the area graze cattle, sheep, and horses.

The cleanup was expected to require several years and, according to an Environment News Service report, "will place the contaminated soil in a lined, capped facility employing the most stringent standards in the country. When complete, the cleanup will allow unrestricted surface use of the mine site for grazing and housing, according to the EPA." People in the area have been at risk "from inhaling radium-contaminated dust particles and radon gas or utilizing contaminated rainwater and runoff that has pooled in the ponds. There is an elevated risk associated with livestock that may graze and [drink] water on the site." Radium exposure at high levels over several years, according to the EPA, "can result in anemia, cataracts, and cancer, especially bone cancer, and death" ("Cleanup" 2011).

"This is an important milestone in the effort to address the toxic legacy of historic uranium mining on the Navajo Nation," said Jared Blumenfeld, administrator for the Pacific Southwest Region. "This plan is the result of several years of collaboration between EPA, the Navajo Nation, and the Red Water Pond Road community living near the mine." "On behalf of the Navajo Nation, I appreciate the efforts of the US EPA and Navajo EPA, and the cooperation from the state of New Mexico to clean up contaminated Navajo trust lands," said Ben Shelly, president of the Navajo Nation ("Cleanup" 2011).

Some other abandoned uranium mines also were being slowly cleansed, but only a small fraction of the several hundred were left idle. One was the Skyline Mine. Jason Musante, an EPA coordinator for a cleanup costing $6 million, told Judsy Fahys of the *Salt Lake Tribune*, "I've got to make this hazard go away as soon as possible. . . . It's already been too long." Radioactive debris was being packed into plastic bins (which Musante described as "giant Tupperware"), then buried underground (Fahys 2011). Most of the mining companies that abandoned sites on Navajo land have gone out of business. General Electric, however, has been billed $44 million for its role in the Northeast Church Rock Mine near Gallup. The EPA itself has spent $60 million. Chevron paid an undisclosed amount to clean up the Mariano Lake Mine in New Mexico. "The government can't afford it; that's a big reason why it hasn't stepped in and done more," said Bob Darr, a spokesman for the Department of Energy. "The contamination problem is vast" (MacMillan 2012). Cleaning all the mines would probably cost hundreds of millions of dollars, according to Clancy Tenley, a senior EPA official who oversees its "uranium legacy program."

Many "hot spots" remained. During the summer of 2010, a Navajo cattle rancher, Larry Gordy, found one such site in his grazing land near Cameron, Arizona, 60 miles east of the Grand Canyon, and called the EPA, which found a radioactivity level that, in two days, would expose a human being (or a sheep) to levels of toxicity considered unsafe for a year. Such exposure can lead to malignant tumors and other serious health problems, according to Lee Greer, a biologist at La Sierra University in Riverside, California (MacMillan 2012).

"If this level of radioactivity were found in a middle-class suburb, the response would be immediate and aggressive," said Doug Brugge, a public health professor at Tufts University Medical School and an expert on uranium's effects. "The site is remote, but there are obviously people spending time on it. Don't they deserve some concern?" (MacMillan 2012).

Elsie Mae Begay's family experienced levels of radiation more than 80 times the EPA limit, as described in the film *The Return of Navajo Boy* (2000). Begay was forced to leave her hogan during cleanup but expressed a strong desire to return home even as Musante admitted that the land "can never go back to what it was before" ("Cleaning Up Uranium" 2011). In 2011, according to the Indian Country Today Media Network, "the Navajo Nation is already dealing with contamination from previous uranium mines and its attendant high rates of cancer, heart disease and birth defects. Cleanup efforts are taking years, and the U.S. Environmental Protection Agency (EPA) is evaluating more than 500 sites in the western part of the Navajo Nation."

Uranium tailings cleanup has its own perils. Parts of the Navajo reservation have been so polluted by uranium mining that families whose members have lived there for hundreds of years are being forced to leave their homes and lands so that contaminated soil can be removed. The mines and mills have been closed, but the waste piles endure, polluting water when it rains and creating deadly dust during long, droughty periods. Federal environmental officials have been stunned at the extent of uranium contamination. Unrecorded mine sites continue to surface, in many cases after people have fallen into abandoned shafts. "It is shocking—it's all over the reservation," said Jared Blumenfeld, the EPA's regional administrator for the Pacific Southwest. "I think everyone, even the Navajos themselves, have been shocked about the number of mines that were both active and abandoned" (Frosch 2014, A-12).

In 2014, Bertha Nez, who lives near Church Rock, was facing the prospect of having to abandon her ancestral home at age 67. She had spent months in Gallup motel rooms as radioactive soil was removed from her community. "This is where we're used to being, traditionally, culturally," she said. "Nobody told us it was unsafe. Nobody warned us we would be living all this time with this risk." An enormous pile of uranium mine waste near Nez's home may require eight years of cleanup. Relocations are voluntary, but many Navajos are skeptical whether radioactivity can be restored to safe levels. "Our umbilical cords are buried here, our children's umbilical cords are buried here. It's like a homing device," said Tony Hood, 64, who worked in the mines. "This is our connection to Mother Earth. We were born here. We will come back here eventually" (Frosch 2014, A-12).

Between 2008 and 2012, U.S. federal agencies spent $100 million on the cleanup, according to the EPA. An additional $17 million was been spent by energy companies responsible for some of the tailings. Waste removal is encumbered by bureaucracy. An application to remove tailings requires paperwork for the NRC, which conducts environmental reviews. Required public hearings also can stall actual removal of toxic waste for several years. "That time frame seems unreasonably long for tribal members, who said that spending so long living away from the reservation has been difficult" (Frosch 2014, A-12). By the end of 2013, the EPA had spent $1 million on temporary housing for residents of Nez's community, much of it billed to General Electric, which acquired the old Northeast Church Rock Mine site in 1997 and its subsidiary, United Nuclear Corporation, operator of the mine.

In a decision handed down on December 12, 2013, Allan L. Gropper, U.S. bankruptcy judge, said that Anadarko and Kerr-McGee are liable for damages between $4.1 billion and $5.1 billion to a large group of plaintiffs that include "the United States, 22 states, four environmental response trusts and a trust for the benefit of certain tort plaintiffs" ("Navajo Nation" 2013). The Navajo Nation was scheduled to receive about $1 billion of the settlement to clean up 49 mining and milling sites. As large as the settlement seemed, it left untouched about 90 percent of the more than 500 abandoned radioactive sites that affect the Navajo.

Navajo Nation President Ben Shelly said that "any funds resulting from this lawsuit are welcomed and long overdue." The Navajo Nation's abandoned uranium mines are part of roughly 2,000 sites across the United States that have been "contaminated with various toxic compounds, radioactive waste and other chemicals that are compromising health for numerous communities. The plaintiffs are joined under a litigation trust, which was seeking $25 billion to clean up the U.S. sites" ("Navajo Nation" 2013). Given the nature of the proceeding (bankruptcy) and the fact that it will be appealed, whether and when the Navajos and other parties will receive anything remains an open question.

The Navajos said that, even lacking payment, legal validation had some value: "While we recognize the uncertainties of the appeal process and the long road that may be ahead of us, this is still a day of celebration for the Navajo Nation. A federal judge has issued a ruling that could result in over a billion dollars being made available for cleaning up some of the uranium contamination from past uranium mining and processing on the Navajo Nation" ("Navajo Nation" 2013).

## *THE LAGUNA PUEBLO AND ANACONDA'S JACKPILE URANIUM MINE*

Until 1982, the Laguna Pueblo (50 miles east of the Navajo Nation in New Mexico) was the site of the world's largest uranium strip mine. The 7,000-acre Anaconda Jackpile mine opened in 1952 and ceased operation when the ore that was available at market prices had been taken. Until it closed, the mine made the Laguna Pueblo people affluent by most reservations' standards. The price of this affluence, in many cases, was slow death by pervasive radiation. When it closed, Jackpile was employing about 800 people. When miners were diagnosed with lung cancer in unusually large numbers, the mine's supporters said what Navajo miners also heard: they had smoked too many cigarettes. John Redhouse of the Southwest Indigenous Uranium Forum disagreed. "The social costs and health impacts outweigh any jobs and money that goes to Laguna," he said. "Whatever apparent benefits accrue do not necessarily go to the communities but to the multinational energy companies" (Knight 2013).

Decades into the Jackpile mine's active life, some people at Laguna knew they were living on borrowed time. One of them was the internationally renowned writer Paula Gunn Allen. Barbara Alice Mann, an honors professor at the University of Toledo and a friend of Allen's, said that "Paula Gunn Allen was wiped out by the same cancer effects, and even before she was diagnosed, she told me that she knew she would die of cancer, simply because of where she grew up and where her family still lived" (personal communication, July 25, 2013).

The agricultural valley that had sustained the Laguna Pueblo for centuries before the advent of the mine also was ruined. Before 1952, the Rio Paguate had provided water for a valley filled with farms and ranches. Winona LaDuke, writing in *Native Americas*, described the post-mine landscape: "Rio Paguate now runs through the remnants of the strip mine, emerging on the other side a fluorescent green in color" (LaDuke 1992, 58). Since 1973, the EPA had known that the mine was leaching radiation into the Laguna Pueblo's water supply. By 1975, the EPA had issued reports describing large-scale groundwater contamination. The Laguna tribal center, the community center, and many reservation houses also were dangerously polluted. The problem was compounded when Anaconda used low-grade uranium ore to surface many roads on the reservation.

By 2013, Jackpile was being cleaned up by the Laguna tribe at a cost of $48 million, paid by ARCO. Many Laguna said that although the "remediation" would restore the mine site to its former condition, uranium already had leached into soil and water, making total cleansing impossible. "Two tributaries near the mine and the Rio San Jose have already tested positive for radiation contamination," said Manual Pino of the Laguna-Acoma Coalition for a Safe Environment. "It's one of the best kept secrets of the United States" (Knight 2013).

Realizing that "Native Americans in the United States and Canada have inherited a grim legacy of increased rates of cancer and a ruined environment because of uranium mining on tribal homelands" (Knight 2013), representatives of several indigenous communities gathered on the Laguna Indian Reservation during June 2013 for the tenth annual conference of the Indigenous Environment Network. As they met, long-shuttered uranium mines in some parts of the United States were being opened again, in some cases to fuel nuclear power plants as an "environmentally friendly" alternative to fossil fuels.

New projects threaten scarce water supplies on Navajo land, said Kathleen Tsosie, secretary of the Eastern Navajo Diné Against Uranium Mining. In Tsosie's hometown, Crownpoint, New Mexico, Hydro Resources Inc. has announced plans "to leach uranium from the groundwater in three places in the Northwestern part of the state. The underground leach mining process is different from traditional open-pit mines since it occurs in the groundwater itself when chemicals are injected into the aquifer to dissolve the ore and is then pumped out. 'How can this not possibly threaten our water supply,' said Tsosie. 'And many of our sacred sites are near these wells' " (Knight 2013). Several Navajo groups petitioned the NRC to revoke the company's mining permit.

"They said the mine would make us rich but I'm still poor and almost everyone around me is dying of cancer and strange diseases," said Dorothy Purley, a woman dying of lymphoma cancer who lived about 1,000 yards from the mine. She also worked for Anaconda Jackpile for eight years (Knight 2013) but was never told that her life was at risk from radiation or what precautions were required to shield herself. In Paguate, a small town where Purley lived her entire life, about 50 miners died from cancers or related maladies. About 20 other people who were not miners lived downwind of the mine also had died by 2013, she said.

Purley said that in one instance,

> When the crusher got stuck we were down at the bottom using hammers to break the rocks down so the crusher could start going back up into processing, and I helped haul the ore from P9 where they were mining underground the richest ore that was coming out. We hauled back to the ore cars to have it transported. We came home, went to bed, never was a word said that there was contamination [because] of radiation. [I] just realized when I came to deal with this cancer, I found out. This cancer has really taught me a lot . . . and gotten me to a lot of places and I've realized what cancer really means. ("Uranium Mining" 1996)

The ore was loaded onto open cars by the Santa Fe Railroad as it blew over everything on its way out of town. No one tested the tracks for radioactive contamination. The mine's crusher was on the east side of town, and the prevailing east wind blew yellow powder all over the area.

Purley was one of few women hired at the mine. She worked in several capacities, from receptionist to security guard, to jobs near the mines. "They used to blast sometimes four times a day," she recalled. "When there was no wind, that sulfur that they put in the blasting powder would just kind of sit over the village and sometimes we had to cover our food. We . . . Indians dried our meat and some of our vegetables out in the sun; we never realized how much contamination there was in the air 'til we realized cancer was the main thing killing our people here in Paguate" ("Uranium Mining" 1996).

Three times, she miscarried. The loss of children broke her heart. Once, she required surgery after nearly bleeding to death. She knew several women who did not miscarry, but their babies were deformed. Once, she watched a cat give birth to kittens with no tails. Many children grew up mentally impaired, and nearly everyone suffered allergies that had been rare or unknown before the strip mine unearthed the yellow powder. Purley herself had asthma and bronchitis before she was diagnosed with breast cancer. People didn't complain because the paychecks put bread and butter on the table. "Money," said Purley, "is the maker of all evil" ("Uranium Mining" 1996).

## THE NORTHWEST TERRITORY DENE: DECIMATED BY URANIUM MINING

At the dawn of the nuclear age, Paul Baton and more than 30 other Dene hunters and trappers who were recruited to mine uranium called it the "money rock." Paid $3 a day by their employers, the Dene hauled burlap sacks of the grimy ore from one of the world's first uranium mines at Port Radium across the Northwest Territories to Fort McMurray. While mining, "many Dene slept on the ore, ate fish from water contaminated by radioactive tailings and breathed radioactive dust while on the barges, docks and portages. More than a dozen men carried sacks of ore weighing more than 45 kilograms for 12 hours a day, six days a week, four months a year. . . . Children played with the dusty ore at river docks and portage landings. And their women sewed tents from used uranium sacks" (Nikiforuk 1998, A-1).

Since then, according to an account by Andrew Nikiforuk in the *Calgary Herald*, at least 14 Dene who worked at the mine between 1942 and 1960 died of lung, colon, and kidney cancers, according to documents obtained through the Northwest Territories Cancer Registry. The Port Radium mine supplied the uranium to fuel some of the first atomic bombs. Within half a century, uranium mining in northern Canada had left behind more than 120 million tons of radioactive waste, enough to cover the Trans-Canada Highway across Canada two meters deep. By 2000, production of uranium waste from Saskatchewan alone occurred at the rate of over 1 million tons annually (LaDuke 2001).

Dene Cindy Gilday said that uranium mining there beginning during the 1940s devastated her hometown of Deline, near Great Bear Lake. Beginning at that time, young Dene men were hired "to carry uranium in sacks from the mines onto barges. The men had no knowledge of the toxic qualities of their loads. . . . It was the first time people at Great Bear Lake started to die of lung, bone, stomach, brain and skin cancer" (Knight 2013). Since 1975, following a 30-year latency period after uranium mining began in the area, hospitalizations for cancer, birth defects, and circulatory illnesses in northern Saskatchewan had increased between 123 and 600 percent. In other areas impacted by uranium mining, cancers and birth defects have increased, in some cases, to as much as eight times the Canadian national average.

"Before the mine, you never heard of cancer," said Baton, age 83. "Now, lots of people have died of cancer" (Nikiforuk 1998, A-1). Declassified documents have revealed that the U.S. government, which bought the uranium, and the Canadian federal government, at the time the world's largest supplier of uranium outside the United States, withheld health and safety information from the native miners and their families.

While many of the Dene blame uranium mining and its waste products for their increased cancer rates, some Canadian officials compiled statistics indicating only marginal increased mortality from uranium exposure. André Corriveau, chief medical officer of health for the Northwest Territories, noted that high cancer rates among the Dene do not differ significantly from the overall territorial profile. He said that the death rate was skewed upward by high rates of smoking. The Dene, in the meantime, maintain that the fact that almost half the workers in the Port Radium mine (14 of 30) died of lung cancer cannot be explained by smoking alone.

Until his death in 1940, Louis Ayah, one of the North's great aboriginal spiritual leaders, repeatedly warned his people that the waters in Great Bear Lake would turn a foul, dirty yellow. According to "Grandfather," the yellow poison would flow toward the village, recalled Madelaine Bayha, one of a dozen scarfed and skirted "uranium widows" in the village (Nikiforuk 1998, A-1).

The first Dene to die of cancer, or what elders still call "the incurable disease," was Old Man Ferdinand in 1960. He had worked at the mine site as a logger, guide, and stevedore for nearly a decade. "It was Christmastime and he wanted to shake hands with all the people as they came back from hunting," recalled Rene Fumoleau, then an Oblate missionary working in Deline. After saying good-bye to the last family that came in, Ferdinand declared, " 'Well, I guess I shook hands with everyone now,' and he died three hours later" (Nikiforuk 1998, A-1).

According to Nikiforuk's account, others died during the next decade. Victor Dolphus's arm came off when he tried to start an outboard motor. Joe Kenny, a boat pilot, died of colon cancer. His son, Napoleon, a deckhand, died of stomach cancer. The premature death of so many men has not only left many widows but also interrupted the transmission of culture. "In Dene society it is the grandfather who passes on the traditions and now there are too many men with no uncles, fathers or grandfathers to advise them," said Cindy Gilday, Joe Kenny's daughter and chair of Deline Uranium Committee (Nikiforuk 1998, A-1).

"It's the most vicious example of cultural genocide I have ever seen," Gilday said. "And it's in my own home." According to Nikiforuk, "Watching a uranium miner die of a radioactive damaged lung is a job only for the brave." He described Al King, an 82-year-old retired member of the United Steelworkers union in Vancouver, British Columbia, who has held the hands of the dying. King described one retired Port Radium miner whose chest lesions were so bad that they had spread to his femur and exploded it. "They couldn't pump enough morphine into him to keep him from screaming before he died," said King (Nikiforuk 1998, A-1).

In exchange for their labors in the uranium mines, the Dene received a few sacks of flour, lard, and baking powder. "Nobody knew what was going on," recalled Isadore Yukon, who hauled uranium ore for three summers in a row during the 1940s. "Keeping the mine going full blast was the important thing" (Nikiforuk 1998, A-1).

The Dene town of Deline was described by one of its residents as

> practically a village of widows. Most of the men who worked as laborers have died of some form of cancer. The widows, who are traditional women were left to raise their families with no breadwinners, supporters. They were left to

> depend on welfare and other young men for their traditional food source. This village of young men, are the first generation of men in the history of Dene on this lake, to grow up without guidance from their grandfathers, fathers and uncles. This cultural, economic, spiritual, emotional deprivation impact on the community is a threat to the survival of the one and only tribe on Great Bear Lake. (Gilday n.d.)

### *PLANS TO BLAST AN ALASKAN HARBOR WITH NUCLEAR BOMBS*

At times, nuclear weapons have been proposed as construction devices in Indian Country. This is not a parody of Dr. Strangelove. During the 1950s, seeking "peaceful" uses for the weapons, the AEC actually proposed dropping five atomic bombs at Cape Thompson on the North Slope of Alaska to blast a harbor. Project Chariot, as it was called, was part of the agrarian-sounding Project Ploughshare, from the biblical concept of peacemaking, beating swords into ploughshares, a piece of bureaucratic poetry invoking civilian uses for nuclear bombs.

The plans were precise—five bombs with 100 times the explosive power of the first atomic bomb that had targeted Hiroshima. The AEC and other agencies recruited Alaskan political leaders and newspaper publishers, but residents in the Inupait village of Point Hope, about 450 people living 30 miles from the intended target site, rallied against the harbor project as a horrid ecological insanity. "Project Chariot Still On," read a banner headline at the top of the front page of the *Tundra Times*.

The Inupiat fought off the project, but radioactive traces that had been used in preliminary tests later were found at the site. The AEC planted radioactive materials in 15,000 pounds of soil obtained from a Nevada nuclear test site (on Western Shoshone land) around Point Hope during 1962. The Native people of the Point Hope area were never informed of other experiments on their area. The U.S. government tested the effects of radioactive bioaccumulation on caribou, lichen, and humans in the area. One such project was made public in 1992 following a Freedom of Information Act request. The residents demanded and won cleanup by the U.S. Department of Energy more than 30 years after they had unknowingly been used as test subjects with live radioactivity (LaDuke and Cruz 2013, 43).

### *THE PRAIRIE ISLAND INDIAN COMMUNITY AND NUCLEAR WASTE*

The 880-member Prairie Island Indian community in southeastern Minnesota was told that a nuclear waste dump next to their reservation would be temporary, but Xcel Energy, which owns a power plant nearby, had come to stay. Although the Indian community had been told that the dump would close in the 1990s, in 2012 the tribe was told that Xcel wanted to extend the life of its "dry cask" storage facility (which warehouses spent fuel rods) another 40 years. About 200

Native people live on the reservation, and as many as 8,000 people a day visit its Treasure Island Resort and Casino (Shaffer 2012).

"Trust is already an issue for the Prairie Island community," reported the Indian Country Today Media Network, "given incidents like the temporary shutdown of one of the facility's reactors on August 14 [2012], when two diesel generators broke down during a routine monthly maintenance test. . . . Failure of the generators, which are in place to help the plant operate during power outages, did not inspire confidence from the tribe even though Xcel assured tribal members that the outage posed no danger. . . . These assurances . . . are reminders of the fact that the 40-year old Prairie Island Nuclear Generating Plant operating a half-mile from our homes relies on aging technology," the Prairie Island Indian Community said. "To have not just one, but both of the back-up diesel generators fail is very troubling. A failure of the back-up diesel generators can affect all other safety features that rely on the electricity that they generate. The failure of both of them during a routine monthly test is simply not acceptable. . . . Our frustrations continue to grow over the federal government's failure to live up to its responsibility to adequately address the nuclear waste issue and as more of that toxic waste builds up next to our community" ("Prairie Island" 2012).

## *THE WESTERN SHOSHONE: "THE MOST-BOMBED NATION OF EARTH"*

Since 1985, many Western Shoshone people, who call themselves "the most-bombed nation on Earth," have joined with environmentalists to resist nuclear weapons testing in Nevada, northwest of Las Vegas, trespassing on the world's only remaining nuclear weapons proving ground. The test site occupies land that President Harry Truman confiscated from the Western Shoshone for national security purposes during 1951, forcibly relocating 100 Native American families. The seizure of Western Shoshone land for nuclear weapons testing has never been fully tested in court.

Traditional Shoshone medicine man Corbin Harney (1920–2007) "came out from behind the bush," as he says, in 1985, initiating protests of the U.S. government's nuclear weapons testing (Harney 1995). (Harney died on July 10, 2007, at age 87, of non-Hodgkin's lymphoma, a cancer that had metastasized into his bones.) He also has taken hundreds of people (Indian and non-Indian) into sweat lodges to pray against testing. Corbin and other Western Shoshone leaders repeatedly have crossed a gated cattle guard onto the 1,350-square-mile test site. On occasion, they have been joined by large groups of people marching in support of a ban on nuclear testing. The marches, which have included several thousand people, sometimes are held on Mother's Day (for Mother Earth). At some of these marches, hundreds of people have been arrested.

Harney cited evidence that the underground tests leak radiation aboveground. "Down-winders," people who live downwind from nuclear weapons test sites, especially in southern Utah, have, according to Harney, "observed extremely high numbers of cancers, leukemia, and other physical deformities in their population." Harney compared these reports to similar findings among citizens of Kazakhstan (in the former Soviet Union) who lived downwind of a nearby test site. They have reported similarly high levels of cancer, leukemia, and birth deformities, among other illnesses. Protests later closed the test site in Kazakhstan. During 1993, when Harney visited Kazakhstan, he learned that local water had been irreversibly contaminated with dangerous levels of radiation. "You can't drink a glass of water there anymore," he said. "All they had to drink was vodka, cartons of juice, and bottled water imported from Europe" (Harney 1995).

About 700 nuclear tests for Great Britain and the United States have since been conducted at the Nevada Test Site since 1951 on lands belonging to the Western Shoshone, according to the Treaty of Ruby Valley, ratified in 1863. Until 1963, the tests were conducted aboveground, a form of testing that was banned by international treaties during the 1960s; in the 1970s, tests also were limited to a 150-kiloton explosive yield.

The EPA has documented cumulative deposits of plutonium in soil samples more than 100 miles north of the test area. "We are the most-bombed nation in the world," said William Rosse Sr., a Western Shoshone elder. "We've had more than our share of radiation, and now they want to dump more nuclear waste on our land at Yucca Mountain" ("Havasupai Fight" 1992). "We have always been a free tribe, never have we been conquered and never have we sold land to the U.S.," said John Wells of the Western Shoshone National Council. "My people are still suffering from the 828 underground nuclear weapons tests and 105 above-ground nuclear weapons tests at the Nevada Test Site" (Hodge 2001).

## *A VILLAGE OF TENTS AND TIPIS BLOCKS A NUCLEAR DUMP IN CALIFORNIA'S WARD VALLEY*

A village of tents and tipis grew in the desert 22 miles west of Needles, California, during the last half of February 1998 as several hundred Native American and non-Indian environmentalists put their bodies on the line in an attempt to stop construction of a dump for low-level nuclear waste on land they regard as sacred. By the third week of February, roughly 250 people were camped at "ground zero" of Ward Valley, 80 acres of federally owned land designated for the waste site. The encampment prevented soil testing necessary for planning of the dump, which had been proposed to receive waste from hospitals, nuclear plants, and other industries that cannot dispose of radioactive materials in other landfills.

By February 19, occupants of the encampment had defied two sets of federal orders to leave the site and had blocked roads leading to it. A 15-day order that had been issued in late January expired on February 14. A new five-day eviction order issued that day expired on February 19, with protesters still eyeball to eyeball with a circle of Bureau of Land Management (BLM) vehicles that had surrounded their camp. Protesters said that the BLM and other federal agents had been rumbling around the camp in large Land Rovers and flying over it in small aircraft and helicopters at odd hours of the night in an attempt to intimidate the campers and deprive them of sleep. Religious ceremonies continued in the camp amid the glare of headlamps and the drone of aircraft overhead. Some of the protesters had chained themselves together. After complaints (and arrival of news media reporters), the vehicles were withdrawn from the perimeter of the camp.

Steve Lopez, a Fort Mojave native spokesman, said that the Ward Valley is central to the creation stories of many Native American peoples in the area as well as the habitat of endangered tortoise species. "Taking away the land is taking away part of ourselves. They used to use bullets to kill our people off. Now it's radioactive waste," Lopez said (Johansen 1998, 7). Ward Valley also is sacred to many Native peoples in the area because of its proximity to Spirit Mountain, the birthplace of their ancestors.

The protesters included a number of elders affiliated with five Colorado River basin tribes (Fort Mojave, Chemehuevi, Cocopah Quechan, and Colorado River Indians) who, along with the rest of the protesters, refused to move. Instead, people in the encampment sent out appeals for supplies and more visitors, asking, according to a Colorado American Indian Movement (AIM) posting on the Internet, "for the physical presence of anyone willing to travel to Ward Valley and participate in the protection of this sacred land and the people defending it." Donations of food, water, blankets, batteries, and rain gear were requested. Wally Antone of the Colorado River Native Nations Alliance said, "Our ceremonies will continue here and our elders will not move. You will have to drag us out, and I say those words with honor" (Johansen 1998, 7).

On February 18, as the BLM's second deadline was set to expire, leaders of the camp invited five BLM officials to their central fire for a religious ceremony, after which the officials were told that the protesters would not move. Andy Mader of Arizona AIM said via the Internet that all cellular phones in the camp were malfunctioning. Some suspected that the government had shut them down. Meanwhile, supporters of the protest outside the camp had received phone calls indicating that several other AIM chapters were sending people to Ward Valley.

Tom Goldtooth, national director of the Indigenous Environmental Network, reported from the camp via the Internet that the protest had become so vigorous not only because the land is regarded as sacred by Native peoples but also because many southern California and Arizona urban areas rely on the Colorado River for water that could be polluted by the proposed waste dump. The same water is used to irrigate crops in both the United States and Mexico.

A position paper circulated by Save Ward Valley, an environmental coalition, stated that the proposed dump lies above a major aquifer 18 miles from the Colorado River. Furthermore, the report asserted that all six of the presently active nuclear waste dumps in the United States are leaking. U.S. Ecology, the contractor selected for the Ward Valley site, presently operates four of those six dumps, the position paper said.

The coalition also asserted that all the waste buried at such sites is not low level. As much as 90 percent of the radioactivity proposed for burial would come from nuclear power plants, including cesium, strontium, and plutonium, the statement said. Speaking on behalf of the coalition, Goldtooth, who coordinated protests by non-Native supporters, evaluated the protest, which eventually caused plans for the dump to be shelved:

> Incorporating the importance of your traditional ways and the use of the sacred Fire as the foundation for guidance and resistance as we fight for environmental justice and Native rights has proven successful at Ward Valley. I applaud the many non-Natives from the peace movement to the anti-nuclear movement and the global human family that raised the consciousness of the world that the sacred tortoise and the ecology of a desert environment must not be sacrificed anymore by the whims of the nuclear waste industry. I witnessed the coming together of the non-Native supporters and the Colorado River tribal communities and Tribal Nations in an historical moment where everyone agreed to fight together with one mind and one spirit to defend the sacredness of the Mother Earth and to defend the sovereignty of the Fort Mojave Tribal Nation. (Goldtooth 2001)

## *WASHINGTON STATE'S YAKAMAS AND HANFORD'S RADIOACTIVE LEGACY*

The Hanford Nuclear Reservation in eastern Washington State (within the treaty boundary of the Yakama Nation) has become a storage site for roughly 53 million gallons of high-level chemical and radioactive wastes that released irradiated water 30 miles upstream from the Yakama Reservation between 1945 and 1989. The presence of this porous depository in their midst has exposed some people to levels of radiation similar to that of the Chernobyl nuclear accident.

Due to this decades-long bath of radiation dumped directly into the river system, oysters caught at the mouth of the Columbia River were so toxic by the early 1960s that when one Hanford employee ate them and returned to work the following day, he set off the plant's radiation alarm. An investigation revealed that the day before, he had eaten a can of oyster stew contaminated with radioactive zinc. The oysters had been harvested in Willapa Bay, along the Pacific Coast in Washington State, 25 miles north of Astoria.

The 560-square-mile Hanford Nuclear Reservation was established during 1943 on lands traditionally used for hunting, fishing, and gathering by the Yakama and Umatilla. The same area is adjacent to the homeland of the Nez Perce. The Hanford facility produced the plutonium used in the first atom bombs dropped on Hiroshima and Nagasaki. Thus, releases of radioactive materials have been contaminating these peoples, as well as the Coeur d'Alene, Spokane, Colville, Kootenai, and Warm Springs tribes, for many years. In 1986, after disclosure that radiation was secretly released into the air and water from 1944 until January 1971, when the last of Hanford's reactors was closed, Yakama leaders were among the first to call for a thorough study of the danger.

For nearly 30 years, ending in 1971, eight of the nine nuclear reactors at the Hanford complex were cooled by water from the Columbia River. Millions of gallons of water that had been pumped directly through the reactor cores picked up large amounts of nuclear material, making the Columbia downstream the most radioactive river in the world, according to state and federal authorities (Schneider 1990, A-9). In July 1990, a federal panel said that some infants and children in the 1940s absorbed enough radioactive iodine to destroy their thyroid glands and to cause an array of thyroid-related diseases.

Although Hanford has stopped producing plutonium, recently released documents indicate that radioactive materials leaking from storage tanks there have continued to contaminate groundwater in the area. The Indigenous Environmental Network reported that "the Hanford site also has eight nuclear reactors that have ... contaminated the Columbia River on which many of the Tribes depend for basic sustenance" ("Uranium/Nuclear" 2001). Unaware of the contamination, indigenous people collected berries near the Columbia, hunted for eels in its tributaries, and took salmon from its waters.

*New York Times* reporter Keith Schneider (1990) described the Colombia below the Hanford Reservation:

> Dammed in the 1950s below Hanford and developed by industrial companies, the river's water is green and gray now. Salmon runs are much smaller than they were before World War II, and Johnny Jackson, a 59-year-old fisherman ... said some fish he catches were marred by deep, infected welts and growths. ... Documents declassified beginning in 1986 said that "radiation spread to the river's bacteria, algae, mussels, fish, birds and the water used for both irrigation and drinking." (A-9)

Radiation in the river has caused concern at Hanford for decades. In 1954, the situation was reviewed in secret meetings at the Washington, D.C., headquarters of the AEC, which operated the plant. Lewis L. Strauss, the commission's chairman, flew to Hanford in the summer of 1954 and was told that "levels of radioactivity in some fish in the Columbia River, particularly whitefish, were so high that officials were considering closing sport fishing downstream" (Schneider 1990, A-9). Ducks, geese, and crops irrigated with Columbia River water also were said to pose a potential health threat.

The public never was alerted. As Schneider (1990) wrote,

> In a memorandum prepared on Aug. 19, 1954, Dr. Parker urged the government to keep the problem secret because the radiation levels were still within safety guidelines and closing sport fishing would compel the plant to discuss the issue publicly and compromise secrecy. "The public relations impact would be severe," he wrote. (A-9)

Secrets, like radioactivity, are prone to leaks as their sources age—here and elsewhere in Indian Country, where recovery from a deadly radioactive legacy has been slow and painful. The same may be said of toxic chemicals, the subject of the next chapter.

## FURTHER READING

Anquoe, Bunty. 1993. "House Begins Investigating Possible Radiation Exposure." *Indian Country Today*, June 9, A-8.

Arnold, Carrie. 2014. "Once upon a Mine: The Legacy of Uranium on the Navajo Nation." *Environmental Health Perspectives*. http://ehp.niehs.nih.gov/122-A44

Baker, James N. 1990. "Keeping a Deadly Secret: The Feds Knew the Mines Were Radioactive." *Newsweek*, June 18, n.p.

Barry, Tom, and Beth Wood 1978. "Uranium on the Checkerboard: Crisis at Checkpoint." *American Indian Journal*, June, n.p.

Brugge, Doug, Timothy Benally, and Esther Yazzie-Lewis, eds. 2006. *The Navajo People and Uranium Mining*. Albuquerque: University of New Mexico Press.

Churchill, Ward, and Winona LaDuke. 1986. "Native America: The Political Economy of Radioactive Colonialism." *The Insurgent Sociologist* 13, no. 3 (Spring): 51–84.

"Cleaning Up Uranium in the Navajo Nation." 2011. Indian Country Today Media Network, May 31. http://indiancountrytodaymedianetwork.com/article/cleaning-up-uranium-in-the-navajo-nation-36290

"Cleanup Planned for Largest Abandoned Uranium Mine on the Navajo Nation." 2011. Environment News Service, September 29. http://www.ens-newswire.com/ens/sep2011/2011-09-29-092.html

Colomeda, Lori A. 1998. "Indigenous Health." Speech. Brisbane, Australia, September 9. http://www.ldb.org/vl/ai/lori_b98.htm

Csordas, T. J. 1999. "Ritual Healing and the Politics of Identity in Contemporary Navajo Society." *American Ethnologist* 26: 3–23.

Eichstaedt, Peter. 1994. *If You Poison Us: Uranium and American Indians*. Santa Fe, NM: Red Crane Books.

Fahys, Judy. 2011. "Uranium Cleanup Under Way on Navajo Land." *Salt Lake Tribune*, May 25. http://www.sltrib.com/sltrib/politics/51833502-90/begay-cleanup-epa-helen.html.csp

Frosch, Dan. 2014. "Amid Toxic Waste, a Navajo Village Could Lose Its Land." *New York Times*, February 20, A-9, A-12. http://www.nytimes.com/2014/02/20/us/nestled-amid-toxic-waste-a-navajo-village-faces-losing-its-land-forever.html

Gilday, Cindy Kenny. N.d. "A Village of Widows." *Arctic Circle*. http://arcticcircle.uconn.edu/SEEJ/Mining/gilday.html

Goldtooth, Tom. 2001. "Indigenous Environmental Network Statement to Nora Helton, Chairwoman, Fort Mojave Indian Tribe, Other Lower Colorado River Indian Tribal Leaders, Tribal Community Members, Elders and Non-Native Groups and Individuals, in Reference to the Ward Valley Victory Gathering to Celebrate the Defeat of a Proposed Nuclear Waste Dump, Ward Valley, California." February 16. http://www.ienearth.org/ward_valley2.html

Grinde, Donald A., Jr., and Bruce E. Johansen. 1995. *Ecocide of Native America: Environmental Destruction of Indian Lands and Peoples*. Santa Fe, NM: Clear Light.

Harney, Corbin. 1995. *The Way It Is: One Water, One Air, One Earth*. Nevada City, CA: Blue Dolphin Publishing.

"Havasupai Fight to Save Grand Canyon from Uranium Mining." 1992. July 21. http://www.ratical.org/ratville/native/havasuEFN.txt

Hodge, Damon. 2001. "Dumpy Meeting: Final Vegas Hearing on Yucca Mountain Plagued by the D.O.E." *Las Vegas Weekly*, September 11. http://www.lasvegasweekly.com/2001_2/09_13/news_upfront7.html

Ivins, Molly K. 1978. "Doctor Charges U.S. Ignored Uranium Spill." *Hutchinson (Kans.) News*, July 27, n.p.

Johansen, Bruce E. 1995. "Victims of Progress: Navajo Miners" [Review of Eichstaedt, *If You Poison Us*]. *Native Americas* 12, no. 3 (Fall): 60–61.

Johansen, Bruce E. 1997. "The High Cost of Uranium in Navajoland." *Akwesasne Notes New Series* 2, no. 2 (Spring): 10–12. http://ratical.com/radiation/UraniumInNavLand.html

Johansen, Bruce. 1998. "Ward Valley: A 'Win' for Native Elders." *Native Americas* 15, no. 2 (Summer): 7–8.

Knight, Danielle. 2013. "Native Americans Denounce Toxic Legacy." TWN: Third World Network, June 14. http://www.twnside.org.sg/title/legacy-cn.htm

LaDuke, Winona. 1992. "Indigenous Environmental Perspectives: A North American Primer." *Native Americas* 9, no. 2 (Summer): 52–71.

LaDuke, Winona. 2001. "Insider Essays: Our Responsibility." Electnet/Newswire, October 2. http://www.electnet.org/dsp_essay.cfm?intID=28

LaDuke, Winona, and Sean Aaron Cruz. 2013. *The Militarization of Indian Country*. East Lansing: Michigan State University Press.

Lee, Tanya H. 2014. "H-Bomb Guinea Pigs! Natives Suffering Decades after New Mexico Tests." Indian Country Today Media Network, March 5. http://indiancountrytodaymedianetwork.com/2014/03/05/guinea-pigs-indigenous-people-suffering-decades-after-new-mexico-h-bomb-testing-153856?page=0%2C3

MacMillan, Leslie. 2012. "Uranium Mines Dot Navajo Land, Neglected and Still Perilous." *New York Times*, March 31. http://www.nytimes.com/2012/04/01/us/uranium-mines-dot-navajo-land-neglected-and-still-perilous.html

Morgan, F. 2001. "Cultural Perspectives on Radiation: With Harmony They Were Placed." Workshop presented at "Creation of the Natural Order" Teachers' Seminar, Uranium Education Program, Diné College, Navajo Nation, July.

"Navajo Nation Could Get $1 Billion in Damages for Uranium Mess." 2013. Indian Country Today Media Network, December 18. http://indiancountrytodaymedianetwork.com/2013/12/18/navajo-nation-could-get-1-billion-damages-uranium-mess-152782

"Navajos Appeal to Inter-American Commission on Human Rights over Proposed Uranium Mining." 2011. Indian Country Today Media Network, May 16. http://indiancountrytodaymedianetwork.com/article/navajos-appeal-to-inter-american-commission-on-human-rights-over-proposed-uranium-mining-34304

Nikiforuk, Andrew. 1998. "Echoes of the Atomic Age: Cancer Kills Fourteen Aboriginal Uranium Workers." *Calgary Herald*, March 14, A-1, A-4. http://www.ccnr.org/deline_deaths.html

"Once upon a Mine: National Institutes of Health Details Uranium's Toxic Legacy to Navajo." 2014. Indian Country Today Media Network, February 6. http://indiancountrytodaymedianetwork.com/2014/02/06/once-upon-mine-national-institutes-health-details-uraniums-toxic-legacy-navajo-153439

Pasternak, Judy. 2010. *Yellow Dirt: An American Story of a Poisoned Lands and a People Betrayed*. New York: Free Press.

"Prairie Island Indian Community Calls for Permanent Nuclear-Waste Solution." 2012. Indian Country Today Media Network, September 5. http://indiancountry

todaymedianetwork.com/article/prairie-island-indian-community-calls-for-permanent-nuclear-waste-solution-132820

Purdom, Tim. 2001. "Radiation Victims Rally; Marchers Complain about IOUs, Delays in Payments." *Gallup Independent*, March 13. http://www.cia-g.com/~gallpind/3-13-01.html#anchor1

Reese, April. 2011. "Navajo Group to Take Uranium Mine Challenge to Human Rights Commission." *New York Times*, May 11. http://www.nytimes.com/gwire/2011/05/12/12greenwire-navajo-group-to-take-uranium-mine-challenge-to-33718.html

Schneider, Keith. 1990. "Washington Nuclear Plant Poses Risk for Indians." *New York Times*, September 3, A-9.

Shaffer, David. 2012. "Prairie Island Tribe Seeks Action on Nuclear Waste." *Star-Tribune*, September 3. http://www.startribune.com/local/168337166.html

"Uranium Miners, Families Bring Tales of Pain to Washington." 2000. *Arizona Republic*, April 15. http://www.ienearth.org/mining_campaign_1c.html#nov17-2000

"Uranium Mining and the Laguna People." 1996. Dorothy Purley, interviewed by Susan Lee, July 1995, in Paguate Village on the Laguna Pueblo Reservation. *Synthesis/Regeneration* 10 (Spring). http://www.greens.org/s-r/10/10-07.html

"Uranium/Nuclear Issues and Native Communities." 2001. Indigenous Environmental Network. http://www.ienearth.org/nuciss.html

Weaver, Jace, ed. 1996. *Defending Mother Earth: Native American Perspectives on Environmental Justice.* New York: Maryknoll.

Woody, R. L., B. Jack, and V. Bizahaloni. 1981. *Social Impact of the Uranium Industry on Two Navajo Communities.* Tsaile, AZ: Navajo Community College.

# CHAPTER 2

# *Toxic Chemicals: A Plague upon the People*

The toxic legacy of Native North America is pervasive but largely invisible to most of us. Many toxic sites are located in out-of-the-way rural areas largely forgotten by majority America that nonetheless supplied its industries with the rudiments of manufacturing for the better part of a century before being closed and cast aside. As of June 5, 2014, 532 (more than one-third) of 1,322 Superfund sites in the United States were in Indian country (Hansen 2014). What follows is a sample of the more prominent sites.

## *"TOO TOXIC TO CLEAN UP"*

Examples abound from sea to shining sea. To cite only one of many, zinc and lead were mined within the jurisdiction of the Quapaw in Picher, Oklahoma, until 1967, when, according to a retrospective by Terri Hansen on the Indian Country Today Media Network, "mining companies abandoned 14,000 mine shafts, 70 million tons of lead-laced tailings, 36 million tons of mill sand and sludge, as well as contaminated water, leaving residents with high lead levels in blood and tissues. Cancers skyrocketed, and 34 percent of elementary-school students suffered learning disabilities" (Hansen 2013). The area was designated as the Tar Creek Superfund site in 1983, but Picher initially was skipped for remedy and assigned to a special environmental hell: too toxic to clean up. Instead, the federal government offered a rock-bottom buyout that paid people to leave town.

This is not an exceptional case. A catalog of such environmental atrocities would fill a very thick volume. What follows is merely a sampler from across North America, beginning in the Arctic, which, in today's world, is nowhere nearly as pristine as many people believe. In Nunavut, the semisovereign Inuit Nation in northern Canada, it has been snowing poison, as polychlorinated biphenyls (PCBs), dioxins, and other chemicals produced by southern industries are swept northward by prevailing winds. Mothers have been warned not to breast-feed their infants. At Akwesasne, in northernmost New York State and nearby Canada, the fish, laced with the same chemicals, are often inedible, and mothers also have been told not to breast-feed babies (and to avoid eating anything from the St. Lawrence River). The Huicholes of Mexico live in a land laced with toxic chemicals. In all of these locations, Native peoples have organized to restore a livable environment. The Inuit, in particular, have taken an active role in efforts to outlaw the use of these toxins in an international agreement: the Stockholm Convention.

Following negotiation of the Stockholm Convention, which outlaws most of the "Dirty Dozen" persistent organic pollutants (POPs), Inuit activist Sheila Watt-Cloutier evoked tears from some delegates with her note of gratitude on behalf of the Inuit people. The treaty, she said, had "brought us an important step closer to fulfilling the basic human right of every person to live in a world free of toxic contamination. For Inuit and indigenous peoples, this means not only a healthy and secure environment, but also the survival of a people. For that I am grateful. *Nakurmiik*. Thank you" (Cone 2005, 202).

It's a long road back, however. Once they become part of the food chain, POPs are very difficult to dislodge. Following the negotiation of international law to outlaw these pollutants came the decades-long battle to enforce the ban and to rid the Arctic food chain of their effects. POPs tend to become locked in the ecosystem, and as of this writing, Inuit still consume traditional foods and breastfeed their babies with caution.

Travel nearly anywhere in North America, and toxicity has become heritage. In Alaska, the Salt Chuck Mine, a source of copper, gold, palladium, and silver between 1916 and 1941, contaminated the Kasaan, Alaska, harvesting grounds for fish, clams, cockles, crab, and shrimp. For decades, the Native people were unaware that their harvests were saturated with effluvia from mine tailings. Even after the area was declared a Superfund site, Pure Nickel, Inc., sought to reactivate mining in 2012.

In California, the Elem Band of Pomo Indians now are suffering elevated levels of mercury in their bodies due to the Sulfur Bank Mine, a Superfund site that adjoins their colony. Terri Hansen (2014) wrote that "nearby Clear Lake is the most mercury-polluted lake in the world, despite the E.P.A.'s spending about $40 million over two decades trying to keep mercury contamination out of the water. Although the E.P.A. has cleaned some of the soil under Pomo roads and houses, pollution still seeps beneath the earthen dam built by the former mine operator, Bradley Mining Co. For years, Bradley Mining has fought the government's efforts to recoup cleanup costs."

## *"WE FEEL LIKE AN ENDANGERED SPECIES"*

"As we put our babies to our breasts we are feeding them a noxious, toxic cocktail," said Sheila Watt-Cloutier, a grandmother who also has served as president of the Inuit Circumpolar Conference. "When women have to think twice about breast-feeding their babies, surely that must be a wake-up call to the world" (Johansen 2000, 27). Watt-Cloutier was raised in an Inuit community in remote northern Quebec. Unknown to her at that time, toxic chemicals were being absorbed by her body and by those of other Inuit in the Arctic.

During a career as activist and diplomat, Watt-Cloutier ranged between her home in Iqaluit (pronounced "Eehalooeet") capital of the new semisovereign Nunavet Territory to and from Ottawa, Montreal, New York City, and other points south, doing her best to alert the world to toxic poisoning and other perils faced by her people. The Inuit Circumpolar Conference represents the interests of roughly 140,000 Inuit spread around the North Pole from Nunavut (which means "our home" in the Inuktitut language) to Alaska and Russia. Nunavut itself, a territory four times the size of France, has a population of roughly 27,000, 85 percent of whom are Inuit.

Many residents of the temperate zones hold fond stereotypes of a pristine Arctic largely devoid of the human pollution that is now so ubiquitous there. To a tourist with no interest in environmental toxicology, the Inuits' Arctic homeland may seem as pristine as ever during its long, snow-swept winters. Many Inuit still hunt polar bears and seals on the pack ice that surrounds their Arctic-island homelands. To the naked eye, the Arctic still *looks* pristine. In Inuit Country these days, however, it's what you *can't* see that may kill you. Such a scene may look pristine until one realizes that the body fat of polar bears and seals is laced with dioxins and PCBs.

## *A POISONED WORLD*

In 1908, Roald Amundsen, one of the earliest European explorers of the Earth's polar regions, said that Arctic peoples "living absolutely isolated from civilization of any kind are undoubtedly the happiest, healthiest, most honorable and most contented of peoples. . . . My sincerest wishes for our Nechilli Eskimo friends is that civilization may never reach them" (Amundsen 1908, 93). Scarcely more than a century later, the environmental threats facing the Inuit today are entirely outside their historical experience. Welcome to ground zero on the road to environmental apocalypse: a place—and a people—that never asked for any of the curses that industrial societies to the south have brought to them. POPs, such as DDT, dioxins, and PCBs, are multiplying up the food chain of the Inuit as air and ocean currents transport the effluvia of southern industry into polar regions. Geographically, the Arctic could not be in a worse position for toxic pollution as a ring of industry in Russia, Europe, and North America pours pollutants northward.

"We are the miner's canary," said Watt-Cloutier. "It is only a matter of time until everybody will be poisoned by the pollutants that we are creating in this world" (Lamb n.d.). "At times," said Cloutier, "We feel like an endangered species. Our resilience and Inuit spirit and of course the wisdom of this great land that we work so hard to protect gives us back the energy to keep going" (personal communication, March 28, 2001).

Most the chemicals that now afflict the Inuit are synthetic compounds of chlorine; some of them are incredibly toxic. For example, one-millionth of a gram of dioxin will kill a guinea pig. In addition to evidence documenting the spread of POPs through the flora and fauna of the Arctic, studies describe the saturation of the same area by high levels of mercury, lead, and nuclear radiation in fish and game.

POPs have been linked to cancer, birth defects, and other neurological, reproductive, and immune system damage in people and animals. At high levels, these chemicals also damage the central nervous system. Many of them also act as endocrine disrupters, causing deformities in sex organs as well as long-term dysfunction of reproductive systems. POPs also can interfere with the function of the brain and endocrine system by penetrating the placental barrier and scrambling the instructions of the body's naturally produced chemical messengers. The latter tell a fetus how to develop from the womb through puberty; should interference occur, immune, nervous, and reproductive systems may not develop as programmed by the genes inherited by the embryo.

Pesticide residues in the Arctic today may include some used decades ago in the southern United States. Although the United States is several hundreds of kilometers from the Arctic's borders, Nunavut receives up to 82 percent of its dioxins from industries there (Suzuki 2000). Some of the pollution in the Canadian Arctic arrives from as far away as western Europe and Japan. Contaminants can reach the Canadian Arctic from Europe in two weeks.

The Arctic's cold climate also slows decomposition of these toxins, so they persist in the Arctic environment longer than at lower latitudes. The Arctic acts as a cold trap, collecting and maintaining a wide range of industrial pollutants, including PCBs, toxaphene, chlordane, and mercury, according to the Canadian Polar Commission. As a result, "many Inuit have levels of PCBs, DDTs and other persistent organic pollutants in their blood and fatty tissues that are five to ten times greater than the national average in Canada or the United States" (PCB Working Group n.d.).

The generation of POPs has become an issue in Watt-Cloutier's home, Iqaluit, on Baffin Island, where the town dump burns wastes that emit dioxins and furans. The dump's plume provides only a small fraction of the Iqaluit residents' exposure to POPs, but it has become enough of an issue to figure in a three-month shutdown of the dump that caused garbage to pile up in the town. The dump was reopened after local public health authorities warned that the backlogged garbage could spread disease and that "the hazard posed by the rotting piles of garbage outweighed the risks of burning it" (Hill 2001, 5). In 2001, Iqaluit's government was asking residents to separate plastics and metals from garbage that can be burned without adding POPs to the atmosphere.

Dioxins, PCPs, and other toxins accumulate with each succeeding generation in breast-feeding mammals, including the Inuit and many of their food sources. The reproductive cycle of humans and other mammals compounds the toxic effects of these chemicals. Airborne toxic substances are absorbed by plankton and small fish, which are then eaten by dolphins, whales, and other large animals.

The mammals' thick subcutaneous fat stores the hazardous substances, which are transmitted to offspring through breast-feeding. Sea mammals are more vulnerable to this kind of toxicity than land animals, so the levels of chemicals in their bodies can reach exceptionally high levels. In ecosystems, POPs "bioaccumulate" in animals at the top of their respective food chains—in the bodies of large meat eaters, such as marine mammals, polar bears, raptors, and humans. Dolphins, seals, and whales in the northern seas are being contaminated. Large land animals, such as caribou, also are affected. Humans who eat these animals form the top (most polluted) rung on the food chain. At the very top of this hierarchy of toxicology is the Inuit mother breast-feeding her infant.

## *INUIT INFANTS: "A LIVING TEST TUBE FOR IMMUNOLOGISTS"*

Inuit infants have provided "a living test tube for immunologists" (Cone 1996, A-1). Due to their diet of contaminated sea animals and fish, Inuit women's breast milk contains six to seven times the PCB level of women in urban Quebec, according to Quebec government statistics. Their babies have experienced strikingly high rates of meningitis, bronchitis, pneumonia, and other infections compared with other Canadians. One Inuit child out of every four has chronic hearing loss due to infections. "In our studies, there was a marked increase in the incidence of infectious disease among breast-fed babies exposed to a high concentration of contaminants," said Eric Dewailly, a Quebec Public Health Center researcher who was among the first scientists to detect high levels of POPs in the bodies of Inuit mothers and their breast milk.

Dewailly, at the time a Laval University scientist, accidentally discovered that Inuit infants were being heavily contaminated by PCBs. During the mid-1980s, Dewailly had first visited the Inuit seeking a pristine group to use as a baseline with which to compare women in southern Quebec who had PCBs in their breast milk. Instead, Dewailly found that Inuit mothers' PCB levels were several times higher than the Quebec mothers in his study group.

Dewailly and his colleagues (Dewailly et al. 1993, 1994, 2000) then investigated whether exposure to organochlorine is associated with the incidence of infectious diseases in Inuit infants from Nunavut. Dewailly and colleagues reported that serious ear infections were twice as common among Inuit babies whose mothers had higher-than-usual concentrations of toxic chemicals in their breast milk. More than 80 percent of the 118 babies studied in various Nunavut communities had at least one serious ear infection in the first year of life (Calamai 2000).

The three most common contaminants that researchers found in Inuit mothers' breast milk were three pesticides (dieldrin, mirex, and DDE, a derivative of DDT) and two industrial chemicals: PCBs and hexachlorobenzene. The researchers could not pinpoint which specific chemicals were responsible for making the Inuit babies more vulnerable because the effects of chemicals may intensify in combination.

The Arctic Monitoring and Assessment Program, a joint activity of Arctic nations and organizations of indigenous Arctic peoples, said in its study

*Pollution and Human Health* that "PCB blood levels, while highest in Greenland and the eastern Canadian Arctic, were high enough (over 4 micrograms of PCBs per liter of blood) that a proportion of the population would be in a risk range for fetal and childhood development problems" (PCB Working Group n.d.). Because they are not easily broken down or excreted, these compounds remain in the body for months or years.

Born with depleted white blood cells, Inuit children have suffered excessive bouts of diseases, including a 20-fold increase in life-threatening meningitis compared to other Canadian children. These children's immune systems sometimes fail to produce enough antibodies to resist even the usual childhood diseases. Inuit babies also have experienced strikingly high rates of bronchitis, pneumonia, and other infections compared with other Canadians. One Inuit child out of every four has chronic hearing loss due to infections. A study published in the September 12, 1996, issue of the *New England Journal of Medicine* confirmed that children exposed to low levels of PCBs in the womb grow up with low IQs, poor reading comprehension, difficulty paying attention, and memory problems.

According to the Quebec Public Health Center, a concentration of 1,052 parts per billion (ppb) of PCBs has been found in Arctic women's milk fat. This compares to a reading of 7,002 ppb in polar bear fat, 1,002 ppb in whale blubber, 527 ppb in seal blubber, and 152 ppb in fish. By contrast, the U.S. Environmental Protection Agency (EPA) safety standard for edible poultry is 3 ppb and for fish 2 ppb. At 50 ppb, soil is often considered hazardous waste. Research by the Canadian federal Department of Indian and Northern Affairs indicates that Inuit women throughout Nunavut have DDT levels nine times that of the averages in Canadian urban areas. Milk of Inuit women of the eastern Arctic has been measured to contain as much as 1,210 ppb of DDT and its derivative DDE, while milk from women living in southern Canada contains about 170 ppb.

Infants in the North are being developmentally harmed by exposure to several contaminants, among them lead, methylmercury, and PCBs. While IQ tests have long been criticized as culturally biased (favoring white, middle- or upper-class Anglo-American children), indigenous infants' cognitive abilities also are penalized by environmental contamination, beginning in the womb and continuing after birth. Physiological damage due to chemical exposure impedes both biological and psychological development. Each chemical has different effects.

Babies in Arctic Canada are at risk of specific effects on their mental abilities, depending on which contaminants they are exposed to in the womb, according to a new study released in 2014. While all of the contaminants are now linked to neurological effects, each seems to have a different effect on infants. For example, according to the study, "PCBs seemed to impair the babies' ability to recognize things they have seen" ("Contaminants" 2014). A study (Boucher et al. 2014) involving 94 Inuit infants and their mothers from Nunavik in northern Quebec measured prenatal exposure to lead, methylmercury, and PCBs by testing cord blood, followed by standard mental development tests of the babies at 6.5 and 11 months of age, taking part in such activities as retrieving toys from under cloth covers, recognizing photographs, and motor-skills tasks, to identify how growing brains were being damaged. According to Boucher et al. (2014), "Each contaminant was independently associated with impairment of distinct aspects of cognitive function with long-term implications for cognitive development—PCBs with visual recognition memory, methylmercury with working memory and an early precursor of executive function, lead with processing speed—deficits that can already be detected during the first year of life." Many of the same scientists have been building their cases regarding the effect of contaminants on IQs for more than two decades. This trend is worldwide: "Similar findings have been reported for children in Michigan, Oswego, N.Y., the Faroe Islands and Taiwan—all areas where many babies are highly exposed to PCBs or mercury from their mothers' consumption of fish or marine mammals" ("Contaminants" 2014). "The last thing we need at this time is worry about the very country food that nourishes us, spiritually and emotionally, poisoning us," Watt-Cloutier said. "This is not just about contaminants on our plate. This is a whole way of being, a whole cultural heritage that is at stake here for us" (Mofina 2000, A-12). "The process of hunting and fishing, followed by the sharing of food—the communal partaking of animals—is a time-honored ritual that binds us together and links us with our ancestors," said Watt-Cloutier (PCB Working Group n.d.).

In many cases, the Inuit have no practical alternative to "country food." Although grocery stores do business in Canadian Inuit towns and villages today, all the food is flown in. No roads or natural land bridges lead south from the villages. The cost of air freight, compounded by distance, raises the price of a quart of milk in the high Arctic to $5 and that of a battered head of lettuce to $4. A tiny frozen turkey the size of a stewing hen may cost $40 in the Arctic.

## *SOCIOECONOMIC CHANGES IN NUNAVUT*

Nunavut could be characterized as the largest "Indian reservation" in the world. The per capita annual income in Nunavut was about US $8,000 in 2013, which doesn't go far in a place where two liters of milk costs $8 and a loaf of Wonderbread retails for $4. While most of Nunavut has few roads and no railroads (communities are connected by regular airline service), all villages now have television and radio, and most have telephone and the Internet, so images imported from the outside world are never far away.

Widespread poverty, unemployment, crime, substance abuse, and a high suicide rate continue to beset Nunavut, which also faces high unemployment. Iqaluit (Inuktitut for "fishing place"), the capital of Nunavut (situated on southeastern Baffin Island about 1,600 miles north of New York City), contains about 7,000 people, many of whom share big-city problems imported from the outside, including alcoholism, drug abuse, and AIDS. The Royal Canadian Mounted Police detachment at Iqaluit has complained that its members spend so much time rounding up drunks that "it's getting in the way of fighting crime" ("Iqaluit Drums" 2001). Three thousand visits to the local drunk tank have been clocked in a year.

The traditional (usually nomadic) Inuit hunting economy endured largely intact until the 1950s and 1960s, when Canadian authorities forced many people to settle in permanent communities. Many Inuit children were pressured to attend church-run boarding schools away from their families. The schools were designed to assimilate Inuit children into the culture of the *qablunaaq*, the white man. The near collapse of the hunting economy has coincided with a sharp rise in suicides. Less than 50 years ago, suicide was virtually unknown among the Inuit of Canada's Arctic. Suicide now has become one of the leading causes of death in Iqaluit. Among young people, suicide has become by far the leading cause of death. Geela Patterson, adviser to the new territorial government on women's issues and herself a suicide-prevention counselor, took her own life during June 1999. Even in a town with the highest suicide rate in Canada (seven times the Canadian average), where few families have not been touched, Patterson's death sent shock waves through the community.

Most of Nunavut's suicides in recent years have been healthy young men ages 15 to 29. Following the destruction of fur markets after the antifur campaigns in Europe and the United States, unemployment rates in northern villages have risen above 50 percent. "When you reduce the usefulness of men in society, it is bound to have psychological effects," said Simona Arnatsiaq, leader of a Baffin Island women's organization. Susan Aglukark, an Inuit singer who has gained pop-star status in southern Canada, sings of the pain caused by suicide in ballads in English and in her Native Inuktitut. "We are investing more money for suicide prevention," said Paul Okalik, whose brother committed suicide. Rachel Attituq Qitsualik, writing in the *Nunatsiaq News*, a weekly newspaper serving the eastern Arctic, said, "I myself, when younger and miserable, have heard the siren call of suicide as a release from suffering, as has my husband, and others that I know and love. . . . My brother gave in to it" (Brooke 2000).

## *ACCUMULATION OF MERCURY IN THE ARCTIC*

In addition to accumulation of POPs, springtime also brings mercury-laced precipitation to the polar regions. The mercury arrives in much the same manner as dioxins and PCBs via prevailing currents in the ocean and atmosphere. Fossil-fuel combustion releases about 6,500 tons of mercury vapor into the atmosphere every year. "In just five months during the spring each year," according to a report by the Environment News Service, "the northernmost coast of Alaska, a region that includes the Arctic National Wildlife Refuge, receives more than double the amount of mercury that falls on the northeastern United States in a year" ("Mercury Contaminates" 2002). Similar pollution also occurs each year in the Antarctic.

Two studies in the March 15, 2002, edition of *Environmental Science and Technology*, a journal published by the American Chemical Society, presented new information about what some scientists had begun to call the "mercury sunrise," which was first reported in the Arctic during 1998. The first sunrise of spring initiates a series of chemical reactions that dump mercury out of the atmosphere and into the polar snowpack, from which it is released into the food web during the brief polar growing season.

Ralf Ebinghaus, an environmental chemist at the Institute for Coastal Research in Geesthacht, Germany, described the mercury sunrise at the German Neumayer research station in Antarctica based on measurements between January 2000 and January 2001 (Ebinghaus et al. 2002). According to an Environment News Service (ENS) report, Steve Lindberg, a research corporate fellow at the Oak Ridge National Laboratory in Tennessee, "presented the first evidence of how this phenomenon leads to elevated mercury accumulation in snow, as measured at the National Oceanic and Atmospheric Administration's Climate Monitoring and Diagnostic Laboratory in Barrow, Alaska" ("Mercury Contaminates" 2002).

A research team led by Lindberg et al. (2002) wrote that

> our data from Barrow, Alaska, at 71 degrees north, show that rapid, photochemically driven oxidation of boundary-layer Hg0 [gaseous elemental mercury] after polar sunrise, probably by reactive halogens, creates a rapidly depositing species of oxidized gaseous mercury in the remote Arctic troposphere. . . . This mercury accumulates in the snow pack during polar spring at an accelerated rate in a form that is bio-available to bacteria and is released with snowmelt during the summer emergence of the Arctic ecosystem. Evidence suggests that this is a recent phenomenon that may be occurring throughout the earth's polar regions. (1245)

The ENS report described the findings of the Lindberg team that may help explain why mercury levels in Arctic seabirds, seals, and beluga whales have increased over the past 20 years, even as global atmospheric emissions of mercury have declined. The mercury sunrise is said by the scientists to depend on ultraviolet light, open water and active sea ice, all of which have been increasing in polar regions during the last few decades. According to the ENS, the Barrow data indicate that the polar sunrise triggers a series of chemical reactions that convert elemental mercury vapor into a soluble form of oxidized gaseous mercury that accumulates in the polar snowpack. Ebinghaus and Lindberg estimated that by 2001, between 50 and 300 tons of mercury were being dumped from the atmosphere into these polar environments each year.

## *TOXIC POLLUTION OF INUIT LAND BY THE MILITARY*

At the same time that Inuit are being poisoned by imported toxins, they also have discovered that parts of their homelands are laced with toxic "hot spots" left behind by abandoned military installations and mines, all also imports from the industrial South. Several of these hot spots are located at or near the 63 military sites in Canada, Greenland, and Alaska that make up the Distant Early Warning (DEW) system of radar sites. At these sites, according to the Arctic Monitoring and Assessment Program's report *Arctic Pollution Issues: A State of the Arctic*

*Environment*, an estimated 30 tons of PCBs were used, and "an unknown amount has ended up in their landfills" (PCB Working Group n.d.).

Under an agreement reached in 1998, Canadian taxpayers, not the U.S. government, are paying most of the $720 million cleanup bill for 51 decommissioned U.S. military sites across Canada. Cleanup of cancer-causing PCBs, mercury, lead, radioactive materials, and various petroleum by-products is expected to take nearly 30 years. Under the arrangement, the United States was absolved of legal responsibility for environmental damage in Canada in exchange for $150 million in U.S. weapons and other military equipment. The cleanup of all 51 American military sites has revealed pollution that newspaper reports in Canada characterized as "staggering" (Pugiliese 2001, B-1). For example, at Argentia, Newfoundland, 70 miles southeast of St. John's, a large U.S. Navy base that opened in 1941 and closed in 1994 left behind PCBs, heavy metals, and asbestos as well as landfills laced with other hazardous wastes. Waste fuels also have contaminated the water table in the area.

Abandoned DEW sites in the Arctic were contaminated with discarded batteries, antifreeze agents, solvents, paint thinners, PCBs, and lead. According to news accounts, "[Canadian] Defence Department scientists have established that PCBs have leaked from the DEW line sites into surrounding areas as far away as 20 kilometers and, in some cases, the chemicals have been absorbed by plant and animal life" (Pugiliese 2001, B-1). Many of the DEW line locations were established in areas where native people hunt and fish. Alaska Community Action on Toxics works with indigenous communities that face toxic contamination from Cold War sites, including the Yu'pik community on St. Lawrence Island. Alaska Community Action on Toxics also has provided the first comprehensive map of more than 2,000 hazardous waste sites in Alaska.

The Inuit also endure pollution from the European and Asian side of the Arctic Ocean. Pesticide residues and other pollutants spill into the Arctic Ocean from north-flowing Russian and Siberian rivers. Decaying Russian nuclear submarine installations on the White Sea have polluted the ocean with nuclear waste, including entire reactor cores from scrapped ships. While many of the former Soviet Union's worst polluters have gone out of business, some prosper despite the fact that their effluent is adding to the Canadian Arctic's toxic overload. The worst offender is the Norilsk nickel smelter located in northern Siberia. Traces of heavy metals from Norilsk's industries have been detected in the breast milk of Inuit mothers.

Geoffrey York, a reporter for the *Toronto Globe and Mail*, described the industrial city of Norilsk, population 230,000, as "the world's most polluted Arctic metropolis" (York 2001, A-1):

> Looming at the end of the road is a horizon of massive smokestacks, leaking pipes, rusting metal, gigantic slag heaps, drifting smog, and thousands of denuded trees as lifeless as blackened matchsticks. Inside malodorous smelters, Russian workers wear respirators as they trudge through the hot suffocating air,

> heavy with clouds of dust and gases. . . . Soviet[-era] research in 1988 found that Norilsk Nickel had created a 200-kilometer corridor of dead forests to the southeast of the city. (A-11)

"We, the Inuit, who for a millennia have lived in harmony with, and with great respect for our land and wildlife are now most impacted by outside forces such as POPs and climate change," said Watt-Cloutier. "With so much already on our plate in terms of attempting to reclaim and restructure our lives to gain back some control over our lives, be it, personal, family, institutions, governance systems etc., it can at times be overwhelming" (Watt-Cloutier, personal communication, March 28, 2001).

## *TOXIC CONTAMINATION OF TRADITIONAL INUIT FOODS IN ALASKA AND RUSSIA*

Toxic contamination of traditional foods has become an issue in Alaska as well as in Nunavut. Native delegates told an international conference on Arctic pollution in Anchorage on May 1, 2000, that some Alaska natives are avoiding their traditional foods out of fear that wild fish and game species are contaminated with pesticides, heavy metals, and other toxins.

"I have a son who has quit eating seal meat altogether," said St. Paul Island resident Mike Zacharof, president of the Aleut International Association. The association was formed because indigenous people in western Alaska and Russia are worried about pollution that crosses their countries' boundaries, Zacharof said, adding, "Fear that seal livers may contain mercury has made many islanders wary of eating the staple of their diet, though most still do" (Dobbyn 2000). "Many people in Prince William Sound no longer eat their traditional foods because of the [*Exxon Valdez*] oil spill. This impacts not only our physical well-being but our emotional and spiritual lives as well," said Patricia Cochran, director of the Alaska Native Science Commission. Cochran added, "Natives from every region of Alaska have been noticing more tumors, lesions, spots and sores on land and sea animals" (Dobbyn 2000). Indigenous reports are being compiled in a report for the Alaska Native Science Commission called the Traditional Knowledge and Contaminant Project, started in 1997. The report was funded by the EPA.

Research by the Circumpolar Arctic Monitoring and Assessment Program has revealed high levels of the pesticide DDT in the breast milk of Russian mothers in Arctic regions. Another study, by the University of Alaska Anchorage, indicates that pregnant Alaska Native women who eat subsistence foods may be exposing their fetuses to potentially dangerous pollutants. "There's no question that people are concerned not only about what it's going to do to them but to their unborn children," Cochran said (Dobbyn 2000). The Tanana Chiefs' Conference, a tribal social services agency for the Interior of Alaska, has detected high levels of DDT in salmon.

Because of toxic contamination, Alaskan Inupiat hunters now examine animals they are preparing to butcher more closely. It is called "playing doctor," wrote

David Hulen, who described one hunter as he "slips on a pair of rubber gloves, wipes clean a titanium-blade knife and begins cutting tissue samples from the seal to be sent off and tested for PCBs, DDT and nearly 50 other industrial and agricultural pollutants" (Hulen 1994, A-5). The hunters, in this case, were accompanied by Paul Becker, a Maryland-based scientist from the National Marine Fisheries Service. Becker has been working with native hunters since 1987, collecting samples for a national marine mammal tissue bank. He has trained several groups of villagers in Nome, Barrow, and other settlements in Alaska to take samples as part of their regular hunts. Hulen (1994, A-5) reported that "all of this has caused no small amount of anxiety in the villages, where oil from rendered seal fat is a staple consumed like salad dressing and where people routinely dine on dishes such as dried seal meat, walrus heart and whale steaks."

Charlie Johnson, executive director of the Eskimo Walrus Commission, a Native group that works with the government to help manage Pacific walrus populations, said, "We've seen a huge increase in cancer rates. People hear about these things turning up and wonder if that has anything to do with it. I don't think there's enough known yet to say there's a problem here in Alaska . . . but I think we have to be aware that we could have a problem eventually" (Hulen 1994, A-5). Because of their role in handling body wastes, the kidneys of mammals are especially vulnerable to accumulation of toxic chemicals. Natives continue to eat the kidneys of the bowhead whale, however, because the kidney is one of the best-tasting parts of the animal.

## *SEWARD, ALASKA: DON'T EAT THE REINDEER*

The Seward Peninsula of Alaska has been extensively mined during the past 100 years for cadmium and various lead-bearing ores that are easily absorbed by plants that provide food eaten by ungulates that concentrate in liver, kidney, and muscle tissues. In some cases, health officials have warned local native peoples to avoid eating reindeer, once a dietary staple. In addition, contamination from weapons testing, accidental pollution, or illegal dumping may have found its way into the lichens of northwestern Alaska, thereby accumulating in reindeer and caribou tissue.

The people on the Seward Peninsula live a subsistence lifestyle in which a high percentage of their diet comes from local plants and animals. The incidence of cancer and other diseases appears to be rising among the indigenous people in this region who subsist on contaminated reindeer and other "country food." The people in local villages are particularly concerned that contamination from air pollution, mining operations, and dump sites are concentrating in the tissue of subsistence animals, posing a health risk. The University of Alaska's Reindeer Research Program has detected high levels of cadmium and lead in several species. If similar concentrations were to be found in meat, consumption of 40 to 60 grams of meat per week would exceed the recommended intake rate ("Heavy Metal" 2000).

## DEFORMED BABIES, MERCURY, AND THE GRASSY NARROWS FIRST NATION IN ONTARIO

Deformed babies are being born on the Grassy Narrows First Nation in Ontario decades after the Dryden Chemical Company dumped 10 tons of neurotoxins into the English-Wabigoon River between 1962 and 1970. More than half of the 160 people who have been examined there have shown symptoms of mercury poisoning, 40 percent of whom have Minamata disease, a form of acute, long-term mercury contamination, with seizures, tremors, tunnel vision, and loss of coordination.

"Both the federal and provincial governments need to recognize and effectively address the lasting issue of mercury exposure in First Nation communities along the English Wabigoon River system," Assembly of First Nations National Chief Shawn A-in-chut Atleo said. "The federal and provincial governments' callous disregard for the health and well-being of the Grassy Narrows community is appalling," said Assembly of First Nations Regional Chief Angus Toulouse, speaking for the Chiefs of Ontario, in a statement. "The poverty and ill health currently being experienced by Grassy Narrows' citizens are the direct consequences of unregulated development, and disregarding the community's right to free, prior and informed consent on activities that occur within their traditional territory" ("Mercury Poisoning" 2012).

## TOO FEW INFANT BOYS? BLAME ESTROGEN-BLOCKING CHEMICALS

The mothers of Canada's Aamjiwnaang First Nations community have been giving birth to a very low percentage of boys, and exposure to estrogen-blocking chemicals may be the reason. "While we're far from a conclusive statement, the kinds of health problems they experience—neurodevelopment, skewed sex ratios—are the health effects we would expect from such chemicals and metals," said Niladri Basu, lead author of a study of the matter, who is also an associate professor at Montreal's McGill University. A study in 2005 revealed that only about 35 percent of Aamjiwnaang infants were male (Bienkowski 2013).

The Aamjiwnaang reserve lies about 15 miles from "Chemical Valley" (with more than 50 chemical factories, oil refineries, and other pollution sources) near the U.S.–Canadian border along Lake Huron, east of Michigan, across the St. Clair River, where about 850 of 2,000 tribal members live on the reserve. Four types of PCBs were measured in 42 children and mothers at Aamjiwnaang at levels two to seven times higher than average than among Canadians. Most of the PCBs, which skew sex ratios, are common in fish in that area, but few of the people had eaten them. Basu suspected that the chemicals remained in the soil and air from decades ago, before they were banned in the 1970s. "Aamjiwnaang" means "at the spawning stream," describing their historical diet of fish, but anyone in the area has been warned since the 1970s not to eat fish from the lake and rivers.

"I'm struck by the elevated PCB levels," said Nancy Langston, a professor of environmental history at Michigan Technological University who was not involved in the study, "and the fact that all are anti-estrogenic." Some PCBs are antiestrogenic, which means that they block the hormone, while others are estrogenic, which means that they mimic it (Bienkowski 2013).

Other health problems are common in the community:

> In addition to PCBs, the mothers and children had elevated levels of cadmium, some perfluorinated chemicals and the pesticides hexachlorocyclohexane and DDT compared with the Canadian average. . . . In addition to the skewed sex ratio, 23 percent of Aamjiwnaang children have learning or behavioral difficulties—a rate about six times higher than children in a neighboring county, according to a 2005 community study. . . . The asthma rate for children on the reservation is about 2.5 times higher than the rest of the county, according to a 2007 study by Ecojustice, a Canadian environmental organization. Birth complications also are commonplace. Of 132 women surveyed in the community in 2005, 39 percent had at least one stillbirth or miscarriage. The average for U.S. women is 15 percent, according to the National Institutes of Health.

Industrialization in the area is increasing as the Sarnia refinery owned by Suncor Energy expands to handle oil sands production from Alberta. This is true in other areas experiencing increases in petrochemical refining: "Men in Fort Saskatchewan—downwind of refineries and chemical manufacturers and oil sands processors—suffer from leukemia and non-Hodgkin lymphoma at higher rates than neighboring communities. . . . Two known carcinogens—butadiene and benzene—were found at higher levels in rural Fort Saskatchewan than in many of the world's most polluted cities," reported Isobel Simpson, lead author of the study and a chemist at the University of California, Irvine (Bienkowski 2013).

## *THE OUJE-BOUGOUMOU CREE ENDURE SEVERE METAL CONTAMINATION*

Ouje-Bougoumou, a Cree community in northern Quebec about 350 miles north of Montreal, has experienced severe metal contamination in its food and water, a decades-old legacy of pollution related to logging and mining. During the early 1990s, the Canadian federal government created a new community for the Ouje-Bougoumou Cree Nation and forced its members to relocate. The move hasn't solved the pollution problem in fish, their main food source.

"I was kind of shocked," said Cree Margo Miascum Cooper. She fears that the water pollutants killed her father, Albert Cooper, who died of cancer in 2000. "I was really concerned, especially for my family" ("Northern Quebec" 2001). Cooper said that the fish that Albert caught and consumed were deformed by pollution from a nearby gold and copper mine. The fishes' eyes were covered in scabs; many of them were missing fins.

A report commissioned by the Grand Council of the Cree examined soil and water samples that revealed unhealthy levels of toxic heavy metals, which were also found in hair samples of nearly two dozen Cree in the same area. The Cree samples were three times higher than average for selenium, four times higher for lead, and five times higher for aluminum. Hair samples also revealed unusually high levels of arsenic and cyanide. "The levels we have observed particularly of cyanide in fish are killing people and will continue to kill people until something is done," said the report's coauthor, Chris Covel ("Northern Quebec" 2001).

In October 2001, the government of Quebec admitted it knew that the Cree of Ouje-Bougoumou were being poisoned by toxic waste that had fouled their water and poisoned their fish. Government reports documented what the Cree had known for years: the fish they had been catching were deformed, missing eyes and fins, and poisoned by three nearby mines. Covel's study found high levels of cyanide, arsenic, mercury, and other heavy metals in the water that yielded the deformed fish. Environment Minister Andre Boisclair at first tried to dismiss the findings, but on October 25, he admitted that the Ouje-Bougoumou Cree were, in fact, being poisoned. Furthermore, he said that the ministry had known about the severity of the pollution since at least 1999. "There is a problem," he said. "Their results confirm ours, so there is a problem" ("Quebec Admits" 2001).

"They just dumped it [pollution] into the water," said Chief Sam Bossum, who said that his people couldn't get the Environment Ministry to do anything about it, so the Cree commissioned their own study. American researcher Chris Covel sampled mine tailings, lake water and sediment, fish, and human hair. The analysis showed high levels of several toxic chemicals. "The levels that we observed, particularly of cyanide in fish, are killing people," he said. As of early 2002, the government of Quebec had not ordered or funded any actions to remediate the water pollution. Boisclair merely told the Cree "to cut down on their fish intake while he orders more studies" ("Quebec Admits" 2001).

## *FOOT-HIGH MUSHROOMS AND TOXIC MOLD SCAR BLACKFEET HOMES*

Blackfeet reservation families in northern Montana have contracted a number of ailments, some of them serious, from a combination of substandard construction practices and toxic chemicals used in reservation housing. Some of the houses, on rotting wooden foundations, have become home to black mold that covers their walls as well as foot-tall mushrooms in some of their basements. Many of the homes' residents complained of chronic headaches, nosebleeds, asthma (as well as other breathing difficulties), general fatigue, dizziness, kidney problems, and several forms of cancer.

Jamie LaPier, who lived in one of the houses 30 miles east of Browning, Montana, said that the mushrooms growing out of the basement carpet were easy to remove but that toxic mold on the walls of bedrooms was impossible to scrub away. Finally, LaPier sealed the rooms off with duct tape and blocked the door with a bookcase to keep children out, as plumbing leaks fed the mushrooms and mold. Roughly 150 reservation houses constructed during the 1970s and

1980s contained wooden foundations treated with chromated copper arsenate, a toxic wood preservative.

The houses were built with funds from the federal Department of Housing and Urban Development and the Blackfeet Housing Authority. Twenty years later, a department spokesperson said that the structural problems were a tribal, not a federal, responsibility (Seldon 2002, B-1). Most of the wooden foundations were guaranteed for 50 years but lasted barely a decade before floors buckled and sank. Walls bowed as windows and doors popped out of their frames, exposing residents to the extreme cold of Montana winters. Most of the houses are only minimally insulated, making some health problems worse.

## *AKWESASNE: THE LAND OF THE TOXIC TURTLES*

Akwesasne (the Mohawk name for the St. Regis Mohawk reservation that straddles the U.S.–Canadian border in New York State) has become a toxic dumping ground riskier to human health than many urban areas. These environmental circumstances have descended on a people whose whole way of life was once enmeshed with the natural world, a place where the Iroquois origin story says the world took shape on a gigantic turtle's back. In our time, environmental pathologists are found turtles at Akwesasne that qualify as toxic waste.

The Akwesasne reservation was abruptly introduced to industrialism with the coming of the St. Lawrence Seaway during the mid-1950s. Soon after the seaway opened, industry began to proliferate around the Mohawks' homeland. A General Motors foundry opened, followed by aluminum plants and steel mills that provided raw materials and parts for the foundry. Akwesasne has been declared the most polluted Indian reserve in Canada and the largest nonmilitary contamination site in the United States. By the mid-1980s, the Mount Sinai School of Medicine, having studied pollution at Akwesasne, advised residents to eat no more than half a pound of locally caught fish per week. Women who were pregnant or nursing, and children under 15 years of age were told not to eat any local fish at all.

When Ward Stone, a wildlife pathologist for the New York State Department of Environmental Conservation, began examining animals at Akwesasne, he found that the PCBs, insecticides, and other toxins were not being contained in designated dumps. After years of use, the dump sites had leaked, and the toxins had gotten into the food chain of humans and nearly every other species of animal in the area. The Mohawks' traditional economy, based on hunting, fishing, and agriculture, was being poisoned out of existence.

The Mohawks started Stone's environmental tour of Akwesasne with a visit to one of the General Motors waste lagoons, a place called "unnamed tributary cove" on some maps. Stone gave it the name "contaminant cove" because of the amount of toxic pollution in it. One day in 1985, at contaminant cove, the environmental crisis at Akwesasne assumed a whole new foreboding shape. The New York State Department of Conservation caught a female snapping turtle that contained 835 parts per million (ppm) of PCBs. The turtle carries a special significance among the Iroquois, whose creation story describes how the world took shape on a turtle's back. To this day, many Iroquois call North America "Turtle Island."

While no federal standards exist for PCBs in turtles, the federal standard for edible poultry is 3 ppm, or about one-third of 1 percent the concentration in that snapping turtle. The federal standard for edible fish is 2 ppm. In soil, on a dry-weight basis, 50 ppm is considered hazardous waste, so that turtle contained roughly 15 times the concentration of PCBs necessary, by federal standards, to qualify its body as toxic waste. During the fall of 1987, Stone found another snapping turtle, a male, containing 3,067 ppm in its body fat—1,000 times the concentration allowed in domestic chicken and 60 times the minimum standard for hazardous waste. Contamination was lower in female turtles because they shed some of their own contamination by laying eggs, while the males stored more of what they ingested.

In 1985, Stone, working in close cooperation with the Mohawks, found a masked shrew that somehow had managed to survive in spite of a PCB level of 11,522 ppm in its body, the highest concentration that Stone had ever seen in a living creature, 250 times the minimum standard to qualify as hazardous waste. Using these samples and others, Stone and the Mohawks established Akwesasne as one of the worst PCB-pollution sites in North America. In 1986, pregnant women were advised not to eat fish from the St. Lawrence, historically the Mohawks' main source of protein. Until the 1950s, Akwesasne had been home to more than 100 commercial fishermen and about 120 farmers. By 1990, fewer than 10 commercial fishermen and 20 farmers remained.

The EPA released its Superfund cleanup plan for the General Motors foundry during March 1990. The cleanup was estimated to cost $138 million, making the General Motors dumps near Akwesasne the costliest Superfund cleanup job in the United States, number one on the SPA's "most wanted" list as the worst toxic dump in the United States. By 1991, the cost was scaled down to $78 million, but the General Motors dumps were still ranked as the most expensive toxic cleanup.

"We can't try to meet the challenges with the meager resources we have," said Henry Lickers, a Seneca who is employed by the Mohawk Council at Akwesasne. Lickers has been a mentor to today's younger environmentalists at Akwesasne. He also has been a leader in the fight against fluoride emissions from the Reynolds plant. "The next ten years will be a cleanup time for us, even without the money," said Lickers. The destruction of Akwesasne's environment is credited by Lickers with being the catalyst that spawned the Mohawks' deadly battle over high-stakes gambling and smuggling. "A desperation sets in when year after year you see the decimation of the philosophical center of your society," he said (Johansen 1993, 19).

The Mohawks are not alone. Increasingly, restrictive environmental regulations enacted by states and cities are bringing polluters to native reservations. "Indian tribes across America are grappling with some of the worst of its pollution: uranium tailings, chemical lagoons and illegal dumps. Nowhere has it been more troublesome than at . . . Akwesasne," wrote Rupert Tomsho, a reporter for the

*Wall Street Journal* (Tomsho 1990, 1). At Akwesasne, Katsi Cook, a Mohawk midwife who has studied the degree to which mothers' breast milk has been laced with PCBs at Akwesasne, said, "This means that there may be potential exposure to our future generations. The analysis of Mohawk mother's milk shows that our bodies are, in effect, part of the [General Motors] landfill" (LaDuke 1994, 45).

By 2012, Ward Stone had suffered several strokes as the Mohawks at Akwesasne paid tribute to his research. Tom Sakokwenionkwas Porter shared the Akwesasne peoples' appreciation of Stone and his work:

> We would like to take this opportunity to express our utmost gratitude to Ward Stone for all that he has done to help the Mohawk people. He has been a strong advocate for the health of Mohawks, especially at Akwesasne. In the 1980s, a midwife from Akwesasne . . . [Tekatsitsiakwa] Katsi Cook got in touch with Ward Stone and expressed concern about the effects of industrial pollution on the health of the people who reside there. Stone found extremely high levels of PCBs, insecticides and other toxins in area fish and wildlife. His work led to irrefutable proof that the dumping of contaminants by nearby factories was responsible for the high level of PCBs found in mother's milk at Akwesasne. As a result, the people at Akwesasne are benefitting from the awareness of what needs to be done to maintain a healthy environment for the generations to come. Ward Stone has worked tirelessly not only as a strong and dedicated spokesperson for the animals, insects, fish, birds, water, air and Mother Earth, but he has also spoken up for the health and welfare of the people of the Northeast. *Niawenko:wa* (thank you), Ward Stone. ("Mohawks Express" 2012)

Decades after PCBs were banned, young people at Akwesasne still carry them in their bodies at levels double the national average. "Significantly higher levels of PCBs were found among individuals who were breastfed as infants, were first born, or had consumed local fish within the past year," the study found (Gallo et al. 2011). The study "also revealed significantly higher levels in those who had eaten fish within the previous year, in those who were first-born, and in those who were breast-fed" (Hansen 2011). "What this study is saying is that these chemicals are extremely persistent in people," said coauthor Lawrence Schell, a professor at the State University of New York at Albany. "Once you're exposed it's difficult to remove that exposure burden" (Hansen 2011).

The Grasse River Alcoa site at Akwesasne continues to affect Mohawk life. "Though the river has been awaiting cleanup for 20-plus years," reported the Indian Country Today Media Network on January 21, 2013, "The only remediation recorded on the EPA website has been the 1995 removal of about 8,000 pounds of PCBs from the facility" ("Mohawks Say" 2013). Various cleanup options range from simply letting nature take its course to a $1 billion cleanup to be billed to Alcoa. In 2013, the EPA had opted for a $243 million plan to dredge some of the contaminated areas and cap others. The St. Regis Mohawk

(Akwesasne) tribe favors a more expensive option that would dredge 7.2 miles of the Grasse River five feet deep. "Capping is not a permanent remedy, and ice scour is a constant threat to any cap in the Grasse River," Akwesasne Environment Division Director Ken Jock said. "Therefore we do not support the capping of the highly contaminated sediments in main channel. Nobody has any real-world evidence that a cap can withstand a major ice jam and ice scour" ("Mohawk Government" 2012).

Larry Thompson, age 56 in 2011, having tired of waiting for General Motors to clean up a Superfund site next to his family's home at Akwesasne, was arrested in 2011 for taking a backhoe into a toxic landfill on August 11 of that year. He was charged with two misdemeanors and two felonies (second-degree criminal mischief and resisting arrest and reckless endangerment) "after he drove onto the notoriously polluted mound, scooped up contaminated soil and loaded it into railroad cars that were waiting to cart away debris from the GM building that is being torn down in the wake of bankruptcy proceedings." Thompson's wife, Dana Leigh Thompson, told the Indian Country Today Media Network that "she has since received numerous calls from people offering to donate time, backhoes and excavators to continue the protest." "Larry was given this order by the clan mother. Bear Clan Mother," Dana Thompson said ("Mohawk Man Arrested" 2011).

The landfill, known as the General Motors–Central Foundry Division Superfund Site at Massena, New York, is one of the country's most severely contaminated toxic sites. It had been capped with plastic, clay, and soil and planted with grass and trees. Dana Thompson said that "it looks like any bucolic scene in rural New York State. But underneath lies a pile of chemicals and PCBs left over from the plant's heyday" ("Mohawk Man Arrested" 2011). "For 32 years we've been waiting for them to clean it up," Dana Thompson said. "Remove the pile, remove the Superfund site, take all their poison out of here and put it into a secure site," Thompson said. "They call it a cleanup but it's really a cover-up, because they're just covering it up" ("Mohawk Man Arrested" 2011).

## *COLVILLE TRIBES RESIST POLLUTION OF THE COLUMBIA RIVER*

Teck Resources Ltd of British Columbia was taken to court by the Colville Confederated Tribes of Washington State for a 100-year record of polluting the Columbia River and forced to admit in 2012 that runoff from one of the company's smelters had been seeping into the river since 1896. "Trail discharged solid effluents, or slag, and liquid effluent into the Columbia River that came to rest in Washington State, and from that material, hazardous materials [under U.S. environmental laws] were released into the environment," said Dave Godlewski, vice president of environment and public affairs for Teck American. "That's what we've agreed to. We've not talked about the amount of the release. We've not talked about the impacts of those releases. We've just agreed that there has been a release in the U.S." ("Colville Tribes" 2012).

Teck was first sued by the Colvilles in 2004. Pollutants, as much as 400 tons a day, included 250,000 tons of zinc and lead as well as 132,000 tons of other hazardous substances, such as more than 200 tons of mercury, cadmium, and arsenic ("Watershed Heroes" 2013). "Water is at the heart of who we are as a people, and while there are still significant issues that remain to be resolved in this litigation we are pleased with the stipulation, and believe it is a positive step for the tribes and our river," Tribal Chairman John Sirois said. "The Columbia River is both a national treasure, and the cultural and spiritual center for the Colville peoples. We hope that this litigation will move us forward as we seek to clean up and protect the health of the Columbia River for all of the people who enjoy and depend on it. We are committed to do all that we can to improve the watershed both for this generation and for the coming generations" ("Colville Tribes" 2012).

This case also set a precedent: the court struck down a previous doctrine barring liability for a foreign company under U.S. law. The plant, 10 miles north of the U.S. border, was dumping into water that entered U.S. territory. On February 23, 2013, the Sierra Club gave the Colvilles its Watershed Heroes Award at a dinner in Spokane, Washington. "We're very grateful for all the sacrifices you have made," Verner said in introducing John Sirois, chairman of the Colville Tribes. "I'm grateful you relayed the history," Sirois said. "I'm grateful for you honoring all the work of the past councils that really put in the time and effort. We have such a great legal team. There are countless people who played a role in this. It's really a validation of who we are as a people. All along the river, those places are named after our people and where we come from: Okanogan, Chelan, Methow, Eniat, San Poil, Lakes, that is who we are. That is where our people are buried. That is where we're born" ("Watershed Heroes" 2013).

## *THE PENOBSCOTS ENDURE ORGANOCHLORINE CONTAMINATION*

Rebecca Sockbeson, a Penobscot, described "the devastating impact of dioxin in my community" to international negotiators of a treaty to eliminate or ban the most widespread POPs (Sockbeson 1999). She said that her nation of nearly 500 people live on an island in the river, close to seven pulp-and-paper mills. Dioxin was being created as a by-product of the chlorine bleaching process in making paper, discharged from all seven of these mills. Dioxin is a highly potent toxic chemical that may cause cancer and other health problems that is being poured daily into an adjacent river. Sockbeson said that her people have survived on the fish from this river but that "now we are dying from it." She continued,

> Neither dioxin nor cancer is indigenous to the Penobscot people, however they are both pervasive in my tribal community. My people face up to three times the state and national cancer rate, moreover, those that are dying of cancer are dying at younger and younger ages, our reproductive generation. This means that unless you take action to eliminate dioxin and other persistent organic pollutants, there will be no Penobscots living on the island by the end of the next century.

The health and survival of Sockbeson's Penobscot band also is threatened by a choice mothers must make: should they breast-feed their children (imparting superior nutrition) and thus pass on to them their body burdens of PCBs and dioxins?

"With this," she concluded,

> I humbly, respectfully and desperately urge you to draft a treaty that insures the existence of the Penobscot and other indigenous peoples who are so disproportionately impacted by dioxin. That the breast and spoon we feed our babies with is not filled with cancer, diabetes, learning disabilities, and attention deficit [disorder]. We ask that you please at this moment, think of your mothers, daughters, sisters, and grandmothers and find the courage to use this unique world power to take action to eliminate dioxin and POPs creating a promise that I may look upon my great-grand children on our common mother, the earth, as my ancestors look upon me now in this room. *Woliwon*—thank you.

### *OREGON TRIBES FIGHT POLLUTION IN AND NEAR PORTLAND*

Native peoples in Oregon have been resisting "a toxic stew that includes arsenic, cyanide, heavy metals, petroleum, pesticides, dioxins, polychlorinated biphenyls and polycyclic aromatic hydrocarbons. Spanning 12 miles of the Willamette River from the Columbia Slough to the Steel Bridge, the site runs through the heart of the historic port city" (Gibson 2013). The area, at the confluence of the Willamette and Columbia rivers, was first listed as an EPA Superfund site in 2000.

The Native peoples have sought to protect bass and catfish as well as marine animals and fish-eating birds that migrate through or live in Portland's harbor. At risk also are several species of salmon protected by treaties. The health of fish passing through Portland's industrial area has become a major health issue for nearby Native peoples. Kelley Point Park also serves Native peoples as a ceremonial site and a launch site for canoe journeys that involve hundreds of people several times a year, gathering shells, berries, and other foods.

At least 80 companies and landowners have contributed to the pollution, some of them now defunct. The EPA said in 2013 that "cleanup options range from $69 million to $2 billion and include a mix of dredging, capping, in-place or after-dredging treatments and use of innovative technologies" (Gibson 2013). The remaining companies (as the Lower Willamette Group) have recommended the least expensive cleanup options, but the EPA has found that this group underestimates the risks to human health.

### *COEUR D'ALENE DEMAND CLEANUP OF MINING WASTE IN IDAHO*

In October 2001, the EPA proposed a $459 million dollar cleanup of accumulated toxic mining waste in Idaho's Coeur d'Alene River basin. This 1,500-square-mile area was devastated in a flood during 1999, when 1 million pounds of lead flowed into Lake Coeur d'Alene. On another occasion, a million pounds of zinc flowed into the same lake.

The Coeur d'Alene tribe was directly affected by this environmental disaster. The cleanup is expected to require 20 to 30 years at a cost of up to $4 billion. Since 1983, a cleanup has been ongoing near Kellogg, Idaho, which was declared a Superfund site by the EPA. By 2001, more than US $200 million had been spent at this site, most of it to replace contaminated soil in local residents' yards and to cap mine tailings ("Hotspots: Idaho" 2001, 3).

The EPA and the state of Idaho took more than 30 years to finalize a plan worth $635 million to clean up the upper Coeur d'Alene River basin waste sites, as it became one of the largest Superfund sites in the United States. The Bunker Hill Mining and Metallurgical Complex Superfund site, as it is officially known, was first designated in 1983. Mining, dating to the 1880s, released more than 100 million tons of waste into the river system. An EPA report said that "until as late as 1968, tailings were deposited directly into the river. Over time, these wastes have spread throughout more than 160 miles of the Coeur d'Alene and Spokane Rivers, lakes, and floodplains" ("Northwest Tribes and EPA" 2012). "This decision sets forth an ambitious, yet thoughtful and methodical approach to reducing risks from metals, making the Coeur d'Alene Basin an even safer, healthier place to live, work and play," said Dan Opalski, director of the EPA's Seattle Superfund office ("Northwest Tribes and EPA" 2012).

In June 2011, the Hecla Mining Company, the largest silver miner in the United States, agreed to pay $263.4 million plus interest to the U.S. government, the state of Idaho, and the Coeur d'Alene tribe to clean the Bunker Hill Mining and Metallurgical Superfund Site, home to one of the country's largest silver mines from the late 1880s to the late 1960s. The cleanup involves about 100 million tons of mine waste. The lawsuit that resulted in the settlement was initially filed in 1991. Eventual cleanup may cost $3 billion over 50 years or more.

According to the EPA, "About 20 miles of streams are unable to sustain a reproducing fish population, and about 10 miles of tributaries have virtually no aquatic life." Children in the area are exposed to hazardous levels of lead in the drinking water, and "more than 15,000 acres of wildlife habitat contain sediments and soils which are acutely toxic to waterfowl" ("Mining Company" 2011).

"Twenty years ago tribal leaders were convinced that not enough was being done to clean up the Coeur d'Alene Basin following a century of mining activity in the Silver Valley," said Chief J. Allan, chairman of the Coeur d'Alene tribe. "Against all odds, the tribe made an unpopular decision to bring one of the largest superfund lawsuits in our nation's history. . . . The tribe is hopeful that this settlement marks a new chapter in the stewardship of the land we all hold dear. The tribe stands together with the United States, the state of Idaho and Hecla to restore our natural resources while we continue to provide economic prosperity to the region." Roughly three-quarters of the settlement will pay for cleanup begun by state and federal agencies in the mid-1980s (to be concluded in the future). The rest will be "dedicated to the remediation and restoration of natural resources" ("Mining Company" 2011).

### *THE RAMAPOUGH: SUFFERING FORD MOTOR'S PCBs, HEAVY METALS, FREON, ARSENIC AND LEAD*

The Ramapough suffered several premature deaths from toxic waste dumped by the Ford Motor Company for 30 years before filing a class-action lawsuit against both Ford and the EPA in 2006. Ford admitted the dumping but said it was legal at the time. With help from a documentary by Jamie Redford (son of Robert Redford) shown on Home Box Office, the dumping of PCBs, heavy metals, Freon, arsenic, and lead between 1967 and 1971 was publicized. The toxics were dumped by Ford's largest automobile assembly plant, located in Mahwah, New Jersey, "into the veritable backyards of the Ramapough Indian people," wrote Vincent Schilling (2011) for the Indian Country Today Media Network.

The communities' children had played in the paint sludge sitting in the landfill areas, painting themselves with the rainbow of dumped paint, and even ate colorful handmade paint pies. The community suffered from immediate health problems, such as nosebleeds and rashes, and now nearly every home in the community has had a cancer-related death.

### *THE YAQUIS: BORDERS DON'T STOP PESTICIDE CONTAMINATION*

The Yaqui are an indigenous farming people who live and work in and near the Yaqui Valley in Sonora, Mexico, spanning the border between the United States and Mexico. Beginning after World War II, due to lack of available water and financing, many of the Yaquis became unable to support their own farms as they had for centuries. The Yaquis were then forced to lease their lands to outsiders, mainly corporate farmers, who were heavy users of pesticides, herbicides, and fungicides. The use of these chemicals, usually applied by aerial spraying, by tractor, and by hand, brought widespread contamination of land, water, and people.

Concurrently, valley farm operations became mechanized, and irrigation and transport systems were established. The result was a "Green Revolution," with farming becoming big business. Yaqui families from the nearby mountain foothills moved into the valley for employment, while some valley residents moved into the foothills to maintain family-scale farms.

Farmers in the valley reported that two crops a year usually were planted, with pesticides applied as many as 45 times per crop. Compounds include multiple organophosphate and organochlorine mixtures as well as pyrethroids. Thirty-three different compounds were used for the control of cotton pests alone between 1959 and 1990. This list includes DDT, dieldrin, endosulfan, endrin, heptachlor, and parathion-methyl, to name a few examples. As recently as 1986, 163 different pesticide formulations were sold in the southern region of the state of Sonora. Substances banned in the United States, such as lindane and endrin, are readily available to farmers living in the Mexican parts of the valley.

In the valley, pesticide use has long been widespread and continues throughout the year, with little governmental control. Contamination of the resident human population has been documented, with women's breast-milk concentrations of lindane, heptachlor, benzene hexachloride, aldrin, and endrin all above limits of the Food and Agricultural Organization of the United Nations after one month of lactation. During 1990, high levels of multiple pesticides were found in the cord blood of newborns and in breast milk of valley residents. Children are usually breast-fed and then weaned onto household foods.

Household insect sprays were usually applied each day throughout the year in the lowland homes. In contrast, the foothill residents maintained traditional intercropping for pest control in gardens. They usually controlled insects in their homes by swatting them. Most of these people were exposed to pesticides only when the government sprayed DDT each spring to control malaria.

Angel Valencia (2000), a spiritual leader of the Yaqui tribe in Sonora, Mexico, in the village of Potam, described the effects of these chemicals among valley residents. Valencia spoke as a representative of the Arizona-based Yoemem Tekia Foundation, an affiliate of the International Indian Treaty Council:

> I have seen with my own eyes the effects of daily contact with these pesticides—it burns their skin, they lose their fingernails, develop rashes and in some cases they have died as a result of exposure to these poisons. . . . The tragedy of this situation makes me both sad and angry—to think of what has been done to the innocent children who are the future of the Yaqui people. They will not be able to grow and develop, as they deserve to.

During the 1990s, Elizabeth Guillette, an anthropologist and research scientist at the University of Arizona, studied the impacts of pesticide exposure on Yaqui children. The studies of Guillette et al. (1998) confirmed the observations of Valencia, who said that exposure to pesticides had "a serious impact on the health and physical and mental development of the children of our villages" (Valencia 2000). Prior to Guillette's research, researchers at the Technological Institute of Sonora in Obregón, Mexico, had shown that children in Sonora's Yaqui Valley often were born with detectable concentrations of many pesticides in their blood and were exposed again through consumption of their mothers' breast milk.

"I know of no other study that has looked at neurobehavioral impacts—cognition, memory, motor ability—in children exposed to pesticides," said neurotoxicologist David O. Carpenter of the State University of New York at Albany. "The implications here are quite horrendous," he said, because the magnitude of observed changes "is incredible—and may prove irreversible" (Raloff 1998). "Although the children exhibited no obvious symptoms of pesticide poisoning, they're nevertheless being exposed at levels sufficient to cause functional defects," observed pediatrician Philip J. Landrigan of Mount Sinai Medical Center in New York (Raloff 1998).

In Guillette et al.'s (1998) study, children of the agrarian region were compared to children living in the foothills, where pesticide use is minimal. The study selected two groups of four- and five-year-old Yaqui children who resided in the Yaqui Valley of northwestern Mexico. These children shared similar genetic backgrounds, diets, water-mineral contents, cultural patterns, and social behaviors. The major difference was the level of their exposure to pesticides. Guillette adapted a series of motor and cognitive tests into simple games the children could play, including hopping, ball catching, and picture drawing.

The study was constructed in this manner to minimize variables that can affect the outcome of a pesticide study on child growth and development. The population had to meet the requirements of similar genetic origin, living conditions, and related cultural and social values and behaviors, all of which are necessary for comparable study and reference groups.

Guillette had assumed that any differences between the two groups would be subtle. Instead, she recalled, "I was shocked. I couldn't believe what was happening" (Luoma 1999). Luoma continued,

> The lowland children had much greater difficulty catching a ball or dropping a raisin into a bottle cap—both tests of hand-eye coordination. They showed less physical stamina, too. But the most striking difference came when they were asked to draw pictures of a person. . . . Most of the pictures from the foothill children looked like recognizable versions of a person. The pictures from most of the lowland children, on the other hand, were merely random lines, the kind of unintelligible scribbles a toddler might compose. . . . It appeared likely they had suffered some kind of brain damage.

During a follow-up trip in 1998, two years after her initial visit, Guillette found that both groups of children (who at that point were in primary school) had improved in drawing ability. While the lowland children's drawings looked more like people than before, the foothill children were drawing far more detailed images. The lowland youngsters were still evidencing some motor problems, particularly with balance. "Some of these changes might seem minute, but at the very least we're seeing reduced potential," Guillette said. "And I can't help wondering how much these kinds of chemicals are affecting us all" (Luoma 1999).

No differences were found in physical growth patterns of the two groups of children. Functionally, however, Guillette et al. (1998) wrote, "The exposed children demonstrated decreases in stamina, gross and fine eye-hand coordination, 30-minute memory, and the ability to draw a person." Guillette gave children red balloons for successful completion of tasks. "Well over half of the lesser-exposed children could remember the color in the object, and all remembered they were getting a balloon. Close to 18 percent of the exposed children could not remember anything," and only half could remember they were getting a balloon. "It was quite a contrast," she said (Mann 2000, C-9).

Guillette said she noticed that exposed Yaqui children would walk by somebody and strike them without apparent provocation. Otherwise, they tended to sit in groups and do nothing. Foothill children, by contrast, were always busy, engaged in group play. "I'd throw the ball to a group of kids. In the valley, one child would get the ball and just play with it himself," she says. The foothills children played with the ball as a group (Mann 2000, C-9). Yaqui mothers from the valley also reported more problems getting pregnant and higher rates of miscarriages, stillbirths, neonatal deaths, and premature births.

While it has not been possible to determine the Yaquis' exposure to specific pesticides, Guillette said that "we know for sure there has been DDT exposure." The Mexican government "does not know what's being used. The farmer does not give out the information. Pesticides are tied to bank loans, and the banks won't reveal what is being used with certain crops. I just assume everything. The other problem is they get a little of this and a little of that and mix it up. It is very important to remember that the situation is no different agriculturally than what you find in California, the Midwest or the East Coast in the U.S." (Mann 2000, C-9).

"Many of these contaminants have similar reactions in the body," Guillette said. "Many disrupt the endocrine system, which regulates body functions, and that's the main reason I looked at subtle changes. The shift may seem slight, but when they occur within a total society, they can have major implications. To me, the approach should not be treatment of the disease or trying to teach compensation for the deficit but to look at the basic problem of contamination" (Mann 2000, C-9).

The same study added: "Valley children appeared less creative in their play. They roamed the area aimlessly or swam in irrigation canals with minimal group interaction. Some valley children were observed hitting their siblings when they passed by, and they became easily upset or angry with a minor corrective comment by a parent. These aggressive behaviors were not noted in the foothills. . . . Some valley mothers stressed their own frustration in trying to teach their child how to draw" (Guillette et al. 1998).

Concluding her studies, Guillette raised a question that summarized concerns of parents in the lowlands of the pesticide-ridden valley: "Environmental change has placed the children of the agricultural area of the Yaqui valley at a disadvantage for participating in normal childhood activities. Will they remain at risk for functioning as healthy adults?" (Guillette et al. 1998).

## *THE HUICHOLES LIVE WITH PESTICIDES AROUND THE CLOCK*

Exposure to toxic pesticides is one of the greatest risks faced by indigenous migrant workers in Mexico, where tobacco growers and other agricultural companies use them on an industrial scale. Many of the workers are denied safety equipment and access to showers and facilities to wash their clothes after having come into contact with pesticides. In addition, many of the workers live in the same pesticide-laced fields that they harvest, exposing them to contamination around the clock.

During 1993, for example, an estimated 170,000 fieldworker arrived in the valleys of Sinaloa during each planting season. Among these workers, roughly 5,000 have been found to suffer from toxic contamination as a result of the handling of or prolonged exposure to pesticides that are used in cultivation (Diaz-Romo and Salinas-Alvarez n.d.).

Of the 35,000 agricultural laborers who worked in the San Quintin Valley of Baja California during 1996, 70 percent were indigenous people. The majority of the indigenous migrant workers who work in the agroindustrial fields of northern Mexico are Mixtecos, Triquis, and Zapotecs from Oaxaca; Nahuas, Mixtecos, and Tlapenecos from Guerrero; and Purhepechas from Michoacan. According to Estela Guzmán Ayala, women (34 percent) and children under 12 years of age (32 percent) constitute two-thirds of the indigenous labor force in the agricultural regions in northern Mexico (Diaz-Romo and Salinas-Alvarez n.d.).

Ruth Franco, a doctor specializing in work-related health and coordinator of the Program for Day Laborers in Sinaloa, estimated that 25 percent of the 200,000 workers in the Sinaloa valleys during the 1995–1996 cycle were children between the ages of 5 and 14. In the fields where these children and their families work, observers asserted that "thousands of used containers and toxic residues that are generated by the annual use of upwards of 8 million tons of pesticides are criminally disposed of in *ad hoc* trash bins, channels, drains, incinerators, and recycled to storing drinking water" (Diaz-Romo and Salinas-Alvarez n.d.).

One segment of the migrants include between 15,000 and 20,000 Huicholes, who inhabit the mountains of the Sierra Madre Occidental. Approximately 40 percent of all the Huichole families leave their communities in the dry season to find employment in the tobacco fields of the Nayarit coast. During the rainy season, the Huicholes traditionally cultivate a combination of corn, chile, beans, squash, and amaranth. The Mexican government has made the traditional cycle difficult by promoting monocultural planting, distributing hybrid seeds of corn that require the use of pesticides and artificial fertilizers, and replacing the mixed seeds that traditionally were used by the Huicholes and other indigenous agricultural peoples. The use of industrial-scale monocultural agriculture breaks down the indigenous traditions of cooperation while, at the same time, increasing malnutrition and alcoholism. The introduction of herbicides such as Paraquat and 2,4-D gradually destroys communal work, placing the health of cultivators and their families in danger.

A report on the travails of the Huicholes said, "To arrive at the tobacco fields the Huicholes make a journey from the sierras under subhuman conditions, arriving hungry and thirsty. The 'valuable and appreciated' human merchandise includes pregnant women, babies incapable of crying, mute from pain, who have recently been born to malnourished mothers or mothers with tuberculosis. Vulnerable elders and even the 'strong' men arrive at these centers in weak condition" (Diaz-Romo and Salinas-Alvarez n.d.).

Favored workers are given purified water, while the remainder are forced to drink water from irrigation channels that draw from the pesticide-laced Santiago River or local wells that also are contaminated with the chemical cocktail used in the tobacco fields. As they toil in the heat, the workers become drenched with sweat, allowing their bodies to absorb pesticide residues more easily. Nicotine in the tobacco also causes skin irritation and hives, called "green tobacco sickness." Child laborers are particularly susceptible to the effects of pesticides and green tobacco sickness.

The harvesting families often spend the entire day and night in the fields, living and sleeping in boxes or under blankets or sheets of plastic beneath the strings of drying tobacco leaves, further exposing themselves to toxic chemicals and tobacco residues. Most have no potable water, drainage, or latrines. Occasionally, "the Huicholes use the empty pesticide containers to carry their drinking water, without paying notice to the grave dangers that this represents, since the majority cannot read the instructions on the labels which may be written in English" (Diaz-Romo and Salinas-Alvarez n.d.).

## FURTHER READING

Amundsen, Roald. 1908. *The North West Passage*. London: E. P. Dutton & Co.

Bienkowski, Brian. 2013. "Contaminated Tribe: Hormone-Blocking Chemicals Found in First Nation Families." Environmental Health News, November 11. http://www.environmentalhealthnews.org/ehs/news/2013/contaminated-tribe

Blais, Jules M., David W. Schindler, Derek C. G. Muir, Lynda E. Kimpe, David B. Donald, and Bruno Rosenberg. 1998. "Accumulation of Persistent Organochlorine Compounds in Mountains of Western Canada." *Nature* 395 (October 8): 585.

Boucher, O., G. Muckle, J. L. Jacobson, R. C. Carter, M. Kaplan-Estrin, P. Ayotte, É. Dewailly, and S. W. Jacobson. 2014. "Domain-Specific Effects of Prenatal Exposure to PCBs, Mercury, and Lead on Infant Cognition: Results from the Environmental Contaminants and Child Development Study in Nunavik." *Environmental Health Perspectives*. http://dx.doi.org/10.1289/ehp.1206323

Brooke, James. 2000. "Canada's Bleak North Is Fertile Ground for Suicide." *Canadian Aboriginal News*, December 18. http://www.canadianaboriginal.com/health/health21a.htm

Cadbury, Deborah. 1997. *Altering Eden: The Feminization of Nature*. New York: St. Martin's Press.

Calamai, Peter. 2000. "Chemical Fallout Hurts Inuit Babies." *Toronto Star*, March 22. http://irptc.unep.ch/pops/newlayout/press_items.html

"Colville Tribes Savor Teck Resources Court Admission That It Polluted Columbia River." 2012. Indian Country Today Media Network, October 9. http://indiancountrytodaymedianetwork.com/article/colville-tribes-savor-teck-resources-court-admission-that-it-polluted-columbia-river-138899

Cone, Marla. 1996. "Human Immune Systems May Be Pollution Victims." *Los Angeles Times*, May 13, A-1.

Cone, Marla. 2005. *Silent Snow: The Slow Poisoning of the Arctic*. New York: Grove Press.

"Contaminants Have Variety of Effects on Arctic Baby IQs." 2014. Environmental Health News, February 7. http://www.environmentalhealthnews.org/ehs/newscience/2014/Feb/arctic-baby-iqs

Dewailly, E., P. Ayotte, S. Bruneau, S. Gingras, M. Belles-Isles, and R. Roy. 2000. "Susceptibility to Infections and Immune Status in Inuit Infants Exposed to Organochlorines." *Environment Health Perspectives* 108: 205–11.

Dewailly, E., S. Bruneau, C. Laliberte, M. Belles-Iles., J. P. Weber, and R. Roy. 1993. "Breast Milk Contamination by PCB and PCDD/Fs in Arctic Quebec: Preliminary Results on the Immune Status of Inuit Infants." *Organohalogen Compounds* 13: 403–6.

Dewailly E., J. J. Ryan, C. Laliberte, S. Bruneau, J. P. Weber, S. Gringras, and G. Carrier. 1994. "Exposure of Remote Maritime Populations to Coplanar PCBs." *Environmental Health Perspectives* 102, suppl. 1: 205–9.

Diaz-Romo, Patricia, and Samuel Salinas-Alvarez. N.d. "Migrant Workers and Pesticides. A Poisoned Culture: The Case of the Indigenous Huicholes Farm Workers." *Abya Yala News: The Journal of the South and Meso-American Rights Center*. http://saiic.nativeweb.org/ayn/huichol.html

Dobbyn, Paula. 2000. "Contaminated Game Has Natives Worried." *Anchorage Daily News*, May 2. http://www.adn.com

Ebinghaus, Ralf, Hans H. Kock, Christian Temme, Jürgen W. Einax, Astrid G. Löwe, Andreas Richter, John P. Burrows, and William H. Schroeder. 2002. "Antarctic Springtime Depletion of Atmospheric Mercury." *Environmental Science and Technology* 36, no. 6 (March 15): 1238–44.

Gallo, Mia B., Lawrence M. Schell, Anthony P. DeCaprio, and Agnes Jacobs. 2011. "Levels of Persistent Organic Pollutant and Their Predictors among Young Adults." *Chemosphere* 83, no. 10 (May): 1374–82.

Gibson, Leah. 2013. "Oregon Tribes Await Superfund Attention for Portland Harbor Site." Indian Country Today Media Network, February 1. http://indiancountrytodaymedianetwork.com/2013/02/01/oregon-tribes-await-superfund-attention-portland-harbor-site-147374

Guillette, Elizabeth A., Maria Mercedes Meza, Maria Guadalupe Aquilar, Alma Delia Soto, and Idalia Enedina Garcia. 1998. "An Anthropological Approach to the Evaluation of Preschool Children Exposed to Pesticides in Mexico." *Environmental Health Perspectives* 106, no. 6 (June). http://www.anarac.com/elizabeth_guillette.htm

Hansen, Terri. 2011. "Akwesasne Mohawk Youth Are Still at Risk of Industrial Pollutants." Indian Country Today Media Network, June 20. http://indiancountrytodaymedianetwork.com/2011/06/20/akwesasne-mohawk-youth-are-still-risk-industrial-pollutants-39158

Hansen, Terri. 2013. "Major Environmental Disasters in Indian Country." Indian Country Today Media Network, October 8. http://indiancountrytodaymedianetwork.com/2013/10/08/7-major-industrial-environmental-disasters-indian-country-151661

Hansen, Terri. 2014. "532 Superfund Sites in Indian Country." Indian Country Today Media Network, June 17. http://indiancountrytodaymedianetwork.com/2014/06/17/532-superfund-sites-indian-country-155316

"Heavy Metal Levels in Reindeer, Caribou, and Plants of the Seward Peninsula." 2000. Current Research Programs, Reindeer Research Program, University of Alaska Fairbanks, April. http://reindeer.salrm.alaska.edu/research.htm

Hill, Miriam. 2001. "Iqaluit's Waste Woes Won't Go Away; City Sets Up Bins Where Residents Can Dump Plastics, Metal." *Nunatsiag News*, July 27, 5.

"Hotspots: Idaho." 2001. *Drillbits and Tailings* 6, no. 7 (October 31): 3.

Hulen, David. 1994. "Hunt Is On for Pollutant Traces in Bering Sea; Alaska Villagers, Scientists Wonder If Toxic Substances Are Endangering Animals and People Who Eat Them." *Los Angeles Times*, August 15, A-5.

"Iqaluit Drunks Keep Cops' Hands Full." 2001. CBC North/News, March 16. http://north.cbc.ca/cgi-bin/templates/view.cgi?/news/2001/03/16/16iqalcohol

Jacobson, Joseph L., and Sandra W. Jacobson. 1996. "Intellectual Impairment in Children Exposed to Polychlorinated Biphenyls in Utero." *New England Journal of Medicine* 335, no. 11: 783–89.

Johansen, Bruce E. 1993. *Life and Death in Mohawk Country*. Golden, CO: North American Press/Fulcrum.

Johansen, Bruce E. 2000. "Pristine No More: The Arctic, Where Mother's Milk Is Toxic." *The Progressive*, December, 27–29.

LaDuke, Winona. 1994. "Breastmilk, PCBs, and Motherhood: An Interview with Katsi Cook, Mohawk." *Cultural Survival Quarterly* 17, no. 4 (Winter): 43–45.

Lamb, David Michael. N.d. "Toxins in a Fragile Frontier." Transcript. Canadian Broadcasting Corporation News. http://cac.ca/news/indepth/north

Lickers, Henry. 1995. "Guest Essay: The Message Returns." *Akwesasne Notes New Series* 1, no. 1 (April–June): 10–11.

Lindberg, Steve E., Steve Brooks, C.-J. Lin, Karen J. Scott, Matthew S. Landis, Robert K. Stevens, Mike Goodsite, and Andreas Richter. 2002. "Dynamic Oxidation of Gaseous Mercury in the Arctic Troposphere at Polar Sunrise." *Environmental Science and Technology* 36, no. 6 (March 15): 1245–56.

Luoma, Jon R. 1999. "System Failure: The Chemical Revolution Has Ushered in a World of Changes. Many of Them, It's Becoming Clear, Are in Our Bodies." *Mother Jones*, July/August. http://www.motherjones.com/mother_jones/JA99/endocrine/html

Mann, Judy. 2000. "A Cautionary Tale about Pesticides." *Washington Post*, June 2, C-9.

McNeel, Jack. 2013. "Watershed Heroes: Colville Confederated Tribes Win Sierra Club Award for Battling British Columbia Smelter." Indian Country Today Media Network, March 6. http://indiancountrytodaymedianetwork.com/2013/03/06/watershed-heroes-colville-confederated-tribes-win-sierra-club-award-battling-british

"Mercury Contaminates Polar Regions." 2002. Environment News Service, March 21. http://ens-news.com/ens/mar2002/2002L-03-21-09.html#anchor1

"Mercury Poisoning Five Decades Later for Grassy Narrows and White Dog First Nation." 2012. Indian Country Today Media Network, June 12. http://indiancountrytodaymedianetwork.com/article/mercury-poisoning-five-decades-later-for-grassy-narrows-and-white-dog-first-nations-117843

"Mining Company to Pay Coeur d'Alene, State of Idaho and U.S. Government." 2011. Indian Country Today Media Network, June 16. http://indiancountrytodaymedianetwork.com/article/mining-company-to-pay-coeur-d%E2%80%99alene%2C-state-of-idaho-and-u.s.-government-38602

Mofina, Rick. 2000. "Study Pinpoints Dioxin Origins: Cancer-Causing Agents in Arctic Aboriginals' Breast Milk Comes from U.S. and Quebec." *Montreal Gazette*, October 4, A-12.

"Mohawk Man Arrested for Taking Backhoe to Superfund Site." 2011. Indian Country Today Media Network, August 12. http://indiancountrytodaymedianetwork.com/article/mohawk-man-arrested-for-taking-backhoe-to-superfund-site-46891

"Mohawk Government Opposes Grasse River Cleanup Plan." 2012. *Watertown Daily Times*, October 2. http://www.watertowndailytimes.com/article/20121002/NEWS09/710029710

"Mohawks Express Gratitude for Akwesasne Health Advocate." 2012. Indian Country Today Media Network, January 6. http://indiancountrytodaymedianetwork.com/article/mohawk-man-fights-to-remove-toxic-hazardous-waste-from-river-104947

"Mohawks Say EPA Alcoa-Superfund Cleanup Plan Falls Short." 2013. Indian Country Today Media Network, January 21. http://indiancountrytodaymedianetwork.com/2013/01/21/mohawks-say-epa-alcoa-superfund-cleanup-plan-falls-short-147131

"Northern Quebec Cree Poisoned by Mining Pollutants: Report." Canadian Broadcasting Corporation News Online, October 20. http://cbc.ca/cgi-bin/view?/news/2001/10/19/cree_quebec011019

"Northwest Tribes and EPA Agree on Cleanup Plan for Upper Coeur d'Alene River Basin." 2012. Indian Country Today Media Network, September 4. http://indiancountrytodaymedianetwork.com/article/northwest-tribes-and-epa-agree-on-cleanup-plan-for-upper-coeur-d%E2%80%99alene-river-basin-132517

PCB Working Group. N.d. "Communities Respond to PCB Contamination." http://www.ipen.org/circumpolar2.html

"PD 2000 Projects." 2001. Arctic Monitoring and Research—Project Directory, April 11. http://amap.no/pd2000.htm

Pugliese, David. 2001. "An Expensive Farewell to Arms: The U.S. Has Abandoned 51 Military Sites in Canada. Many Are Polluted, and Taxpayers Are Paying Most of the $720 Million Cleanup Cost." *Montreal Gazette*, April 28, B-1.

"Quebec Admits Toxic Waste Poisons Cree." 2001. Canadian Broadcasting Corporation News Online, October 26. http://cbc.ca/cgi-bin/view?/news/2001/10/26/cree_tox011026

Raloff, J. 1998. "Picturing Pesticides' Impacts on Kids." *Science News* 153, no. 23 (June 6):358.

Schilling, Vincent. 2011. "Documentary Featuring the Fight between Ford and Ramapough Indians Premieres Tonight." Indian Country Today Media Network, July 18. http://indiancountrytodaymedianetwork.com/article/documentary-featuring-the-fight-between-ford-and-ramapough-indians-premieres-tonight-43261

Seldon, Ron. 2002. "Families Sicken and Die in Mold-Plagued HUD Housing; Residents Fight Back against Substandard Construction." *Indian Country Today*, April 3, B-1, B-2.

"Silver Valley Cleanup Plan to Cost $635 Million." 2012. *Seattle Times*, August 28. http://seattletimes.com/html/localnews/2019011322_apidsilvervalleycleanup1stldwritethru.html?syndication=rss

Sockbeson, Rebecca. 1999. "Statement by Rebecca Sockbeson for IRATE (Indigenous Resistance Against Tribal Extinction), and I.E.N. (Indigenous Environmental Network)." International POPs Elimination Network, PCB Working Group, September 8. http://www.ipen.org/irate.html

Suzuki, David. 2000. "Science Matters: POP Agreement Needed to Eliminate Toxic Chemicals." December 6. http://www.davidsuzuki.org/Dr_David_Suzuki/Article_Archives/weekly12060002.asp

Tomsho, Rupert. 1990. "Reservations Bear the Brunt of New Pollution." *Wall Street Journal*, November 29, 1.

Valencia, Angel. 2000. "Statement of Angel Valencia, Yoemem Tekia Foundation, Tucson, Arizona POPs Negotiations, March 23, 2000." Native News: The Mail Archive, March 27. http://www.mail-archive.com/natnews@onelist.com/msg00572.html

"Watershed Heroes: Colville Confederated Tribes Win Sierra Club Award for Battling British Columbia Smelter." Indian Country Today Media Network. March 6, 2013. http://indiancountrytodaymedianetwork.com/2013/03/06/watershed-heroes-colville-confederated-tribes-win-sierra-club-award-battling-british

York, Geoffrey. 2001. "Russian City Ravaging Arctic Land." *Toronto Globe and Mail*, July 25, A-1, A-11.

# Chapter 3

## *Canada's Hydroelectric Dreams*

The imaginations of dam builders have only rarely been restrained by the fact that the land they want to flood is owned and occupied by Native Americans. Beginning in the 1970s, Hydro-Quebec, an unwelcomed guest, came with intentions to stay on Cree land in northern Quebec, proposing a vast network of dams that began to transform the Cree homeland, introducing toxic methylmercury. The Cree organized, enlisted support worldwide, and fought off the threat, agreeing later to a scaled-down proposal with economic incentives. In the meantime, the developers were making plans in Manitoba and Labrador.

### *HYDRO-QUEBEC RESHAPES THE CREES' HOMELANDS*

During the 1970s, planners and engineers at Hydro-Quebec indulged themselves in dreams of an electric empire that would make their utility the largest supplier of electricity in the world, spreading into the United States. The utility planned to harness dozens of rivers flowing into James Bay, reshaping the ecology of an area the size of Iowa in the service of electrical generation.

At first, the planners at Hydro-Quebec paid very little attention to what might become of the flora and fauna of the James Bay region during what they envisaged as the largest earthmoving project in the history of humankind. Although it is a public entity (a "Crown corporation" in Canada), Hydro-Quebec found itself unrestrained even by a public board of directors. Hydro-Quebec's electric dreams met problems other than the Crees' resistance. By the time "Phase I" had been completed, by 1990, the utility was $26 billion in debt, mainly due to expenses associated with the planning and construction of its James Bay projects.

The James Bay area is home to 10,000 Cree and 6,000 Inuit people who were not consulted before construction began on "James Bay I." James Bay hydroelectric development was first interjected into the Crees' lives April 30, 1971, when Quebec Premier Robert Bourassa (after whom part of the complex later was named) announced its inception to the people of the province.

The Cree occupy the southern part of the region in Manitoba, Ontario, and northern Quebec as far north as Whapmagoostui (the Cree name for the community of Great Whale). Inuit communities dot the eastern shores of Hudson Bay in Quebec north from Kuujjuarapik to Ivujivik and Salluit. In the Northwest Territories, Inuit communities extend from Arviat on the western shore of Hudson Bay to Coral Harbour on Southampton Island. The Inuit community of Sanikiluaq is located on the Belcher Islands in southeastern Hudson Bay, about 100 kilometers from the mouth of the Great Whale River.

Hunting and fishing make up the heart of the Cree and Inuit economies. The Cree have traditionally hunted migratory birds, particularly during the spring, as well as terrestrial mammals, such as moose. The Cree fish the rivers in the region and trap fur-bearing mammals, such as muskrat and beaver. Traditionally, Inuit harvested fish and marine mammals, such as seals, walrus, and whales. Some communities also depend heavily on caribou.

The Cree and Inuit soon found themselves challenging the construction project in court. It was a matter of life and death not only for traditional cultures but also for the peoples themselves. The construction ruined many traditional hunting areas, and replaced them with poisoned earth laced with life-threatening mercury compounds.

The Cree and Inuit of northern Quebec signed a comprehensive claims agreement (formally named the James Bay and Northern Quebec Agreement) with the Canadian federal and Quebec provincial governments in 1975. The Cree entered agreements (in essence, treaties) with the governments of Quebec and Canada, but they found the terms of these treaties being violated by the ecological consequences of hydroelectric construction.

The 1975 agreement provided the Cree $22 million (Canadian) and the right to continue to hunt, fish, and trap in their usual territories. Not even Hydro-Quebec anticipated that its plans would make a significant portion of the territory unusable for any of those activities. Hydro-Quebec's lawyers were more interested in other stipulations of what came to be called a "treaty" between the Cree and Quebec: the clauses that gave their clients the right to undertake construction of the James Bay project's first phase. In 1979, the first of four projected power stations at La Grande came on line. With its advent, roughly 4,200 square miles of Cree hunting and trapping grounds were flooded. Three hundred black bears drowned as the reservoirs filled. Ignoring natural seasonal cycles, Hydro-Quebec's engineers filled their reservoirs just after the bears had put themselves to sleep for the winter, drowning them. The fact that the black bear is the most sacred animal in Cree mythology did not bother them.

## *TRANSFORMATION OF THE CREES' HOMELAND*

During the James Bay project's first phase, large areas of the Crees' homelands were transformed. Rivers that once spawned large numbers of fish were reduced to trickles or stopped entirely by the creation of reservoirs behind energy-generating dams. Forests were clear-cut and burned, adding greenhouse gases to the atmosphere. More than 10,000 caribou drowned after Hydro-Quebec's rearrangement of the landscape spilled deep water across their migration routes. Hydro-Quebec called the deaths an "act of God" (Biegert 1995, n.p.). This incident was yet another reminder that while hydroelectricity is often touted as "clean" energy that does not emit pollution or directly increase the atmosphere's overload of greenhouse gases, it is not environmentally benign.

Hydro-Quebec's "James Bay II" proposed to dam eight major rivers in northern Quebec at a cost of more than $170 billion to provide electricity to urban Canada in the St. Lawrence River valley and to the northeastern United States. Little is said these days of the original James Bay project's third proposed stage, the most ambitious electric dream of all. As originally envisaged by the engineers of Hydro-Quebec, the project's second phase (which was successfully impeded by the Cree) was to be followed by an even more ambitious "Phase III," a $100 billion proposal to build a 100-mile dike across the mouth of James Bay, "separating it from Hudson Bay so that the now fresh water from James Bay can be pumped (possibly using nuclear-powered pumps) to the Great Lakes and thence to the Midwestern and Southwestern United States" (Native Forest Network 1994). Given free range, the empire builders of Hydro-Quebec were planning to create a utility that would become a power merchant not only to Quebec, Ontario, and New England but also to the eastern half of the United States, with the raw material, electricity, provided by "clean" hydroelectric energy.

Formally called the "Grand Canal," or the "Great Recycling and Northern Development" proposal, Phase III of the James Bay Project would have transformed James Bay into a gigantic freshwater lake. According to Hydro-Quebec's plans, "Fresh water from this reservoir would be diverted south through existing watercourses and new canals to the Great Lakes and thence to the American Midwest and the thirsty, booming Southwest, as well as to the Canadian prairies" (Sierra Club n.d.).

According to now-abandoned plans, once Phase II was completed, Hydro-Quebec estimated that its generating capacity would rise to 27,000 megawatts, as much electricity as could be made available by fully harnessing the power of Niagara Falls (on both sides of the U.S.–Canadian border) 13 times ("Sustainable Development" n.d.). Following construction of the James Bay project's first phase, Hydro-Quebec was generating one-quarter of the electricity consumed in the United States and Canada. Hydro-Quebec engineers estimated that by the time James Bay itself was dammed at the end of Phase III, their utility would be able to meet the demands of two-thirds of the same market.

## *THE TOLL OF METHYLMERCURY*

Just as Hydro-Quebec's electric dreams reached a point of ecocidal fantasy, the Cree and Inuit peoples got in the way. In a very basic way, the Cree were called during the 1980s to protest the James Bay project as a matter of life and death, as they faced methylmercury contamination caused by the earthmoving of the James Bay project's first phase.

Mercury contamination occurs when vast areas of land are disrupted. Plant decay associated with the James Bay project's earthmoving caused large amounts of ordinary mercury to become methylmercury, a bioaccumulative poison. Rotting vegetation accelerated microbial activity that converts elemental mercury in previously submerged glacial rocks to toxic methylmercury. By 1990, many Crees were carrying in their bodies 20 times the level of methylmercury considered safe by the World Health Organization. Hydro-Quebec did not anticipate the accelerated release of methylmercury into the waters of the region, contaminating the entire food chain for the Cree and Inuit as well as birds, fish, and other animals. This type of mercury poisoning can cause loss of vision, numbness of limbs, uncontrollable shaking, and chronic brain damage. By the late 1980s, the Quebec government's health ministry was telling the Cree not to eat fish from their homeland.

Fishing represents more than sustenance for the Cree. Fishing activities are important in knitting family and community. Mercury contamination thus disrupted an entire way of life. Before 1978, concentrations of mercury in 700-millimeter pike were approximately 0.6 microgram per kilogram. After completion of Phase I, the concentrations increased gradually. In 1988, concentrations were 3 milligrams per kilogram, five times the original concentration and six times the maximum permissible concentration for commercial fish (e.g., for human consumption) in Canada (Dumont 1995).

By 1984, the concentration of mercury in the hair of Crees at all ages was much higher than in surveys conducted during the 1970s. During 1993 and 1994, the Cree Board of Health completed an assay of mercury levels in the hair of the Cree. This survey revealed a wide variation in exposure levels between different communities. "If the 6 mg/kg maximum hair concentration recommended by the World Health Organization is used, at least half of the population of several communities is over that limit. In 1984, 98 percent of the population surveyed in the village of Whapmagoostui had mercury concentrations above 6 mg/kg" (Dumont 1995). Mercury levels increased generally with a person's age. Persons recognized as trappers in their communities consistently tested for higher mercury concentrations. "During the past ten years, the number of individuals with high mercury concentrations has decreased considerably" due in large part to programs persuading the Cree to avoid eating tainted fish (Dumont 1995).

The Crees, wrote Charles Dumont, have expressed great concern for the effects of mercury on future generations: "Mercury is presumably toxic at lower doses on the fetus and can cause delays in walking and speaking, and abnormal reflexes. Boys are probably more sensitive than girls. In higher doses it can cause cerebral palsy" (Dumont et al. 1998). Dumont and colleagues examined changes in mercury levels within the Cree population between 1988 and 1993–1994. Hair-sample assays for mercury content were taken by the Cree Board of Health and Social Services during 1988 and again during 1993 and 1994 in all nine Cree communities of northern Quebec. The studies found that

> the proportion of the Cree population with mercury levels in excess of 15.0 mg/kg declined from 14.2 percent in 1988 to 2.7 percent in 1993/94. Wide variations in mercury levels were observed between communities: 0.6 percent and 8.3 percent of the Eastmain and Whapmagoostui communities respectively had mercury levels of 15.0 mg/kg or greater in 1993/94. Logistic regression analyses showed that significantly higher levels of mercury were independently associated with male sex, increasing age and trapper status. There was a correlation between the mercury level of the head of the household and that of the spouse. . . . Mercury levels in the Cree of James Bay have decreased in the recent past. Nevertheless, this decrease in mercury levels may not be permanent and does not necessarily imply that the issue is definitively resolved. (Dumont et al. 1998)

In 1984, two of every three people in Chisasibi, a community of 2,500 at the mouth of the La Grande River, had unhealthy levels of mercury in their bodies; some elders who exhibited symptoms of mercury poisoning had 20 times the usual levels ("Sustainable Development" n.d.).

## *THE CREES ORGANIZE AGAINST JAMES BAY II*

When James Bay II was proposed, the Cree and other Native Americans peoples living in the area not only took the case to court but also sought to organize the customers of the electric utilities that would receive the power, mostly in eastern Canada and New England. This activism ultimately convinced an activist nucleus of customers to pressure their utilities to refuse Hydro-Quebec's power on moral grounds.

The Cree became very effective at addressing public forums in New York and Vermont to tell people that they shared complicity with Hydro-Quebec in the devastation of northern Quebec. The Crees also urged electricity consumers and utilities to conserve and to consider other sources of supply. Several non-Indian environmental groups joined with the Grand Council of the Cree against the James Bay projects. In New York and New England, electricity users were urged to conserve energy to eliminate the need for more generating capacity.

The James Bay projects did not envisage the construction of traditional dams in fast-flowing waterways bordered by mountains, the usual situation in the North American West. The plan was to rework the dozens of rivers that traverse rolling countryside on their way into James Bay into a network of short dams and shallow reservoirs. In 1971, at its inception, Prime Minister Bourassa called Hydro-Quebec's designs for the James Bay region "the project of the century" as he bemoaned the fact that "every day, millions of potential kilowatt hours flow downhill and out to sea. What a waste!" (Biegert 1995, n.p.). Bourassa made his support of Hydro-Quebec a matter of provincial pride and sometimes questioned the patriotism of the project's opponents. He dreamt out loud that northern

Quebec would become a perfect location for magnesium and aluminum smelters that would draw on the newly generated power.

In replies to Bourassa and Hydro-Quebec, the Crees (often through their spokesman, Cree Grand Chief Mathew Coon-Come) asserted that "beavers are the only ones who should be allowed to build dams in our territory" (Biegert 1995, n.p.). The Crees had come to this conclusion as scientists from Quebec's health ministry warned everyone (especially women of childbearing age) not to eat local fish, which were becoming contaminated with methylmercury. Predatory fish, such as pike, had been accumulating a great deal of methylmercury in their bodies since Hydro-Quebec began its ambitious plans to reduce the natural landscape of northern Quebec into a shape and a function suitable for generating electricity. Until then, fish had been a major constituent of the Cree diet for thousands of years.

Coon-Come (2001) said, "Nobody asked us or told us, but three or four times from 1670 on, our people and our lands were handed between kings and companies and countries. We found out about all of this when they came to build the dams and when the courts told us that we did not have any rights, that we were squatters on our own land."

On land, the depredations of the earthmovers are matched by another tendril of life to the south: hunters in orange suits arriving from Canada's urban areas, arriving in late-model sport-utility vehicles along freshly paved roads, who left behind the lifeless, rotting bodies of moose and caribou, their trophy heads removed. Such behavior profoundly offended the Cree.

Coon-Come (2001) recalled,

> In the early 1970s, I was a young student in Montreal when I read in the newspaper that Hydro-Quebec . . . was going to build a hydroelectric mega-project of the century in our territory, by diverting and damming the rivers and flooding our traditional lands. I looked at a map and saw that my family's trap-line and our community was going to be underwater! I immediately returned home, and as a result of a speech I made from the back of a community hall I was launched on my political career.

"Twenty-five years later," Coon-Come recalled, "our people know that the treaty we signed was more broken than honored. Many of the most important promises, such as for economic involvement and development, for protection of the environment and our traditional economy and way of life, and for housing, community development, and infrastructure have been twisted, ignored, or broken" (Coon-Come 2001).

James Bay I involved nine dams, 206 dikes, five major reservoirs, and the diversion of five major rivers over an area roughly the size of Connecticut. Rotting vegetation in the area had released about 184 million tons of carbon dioxide and methane into the atmosphere by 1990, adding to global warming. In the meantime, James Bay I had saddled each of Hydro-Quebec's ratepayers with an average of $3,500 in debt (LaDuke 1993, A-3). James Bay I uprooted 2,000 Cree from Chisasibi Island (also known as Fort George Island). "It's painful to look at young people caught between two worlds. I've seen their faces, knowing that their heritage is under water. It was a whole disintegration of the spirit—all that was drowned," said Coon-Come ("Canadian Indians" 1990, A-2).

In addition to its toll on the Cree, hydroelectric development ruined wetlands and coastal marshes that were important staging grounds for migratory waterfowl, including several species of duck, teal, and goose. The area also was home to rare and endangered species of freshwater seals, beluga whales, polar bears, and walruses. Anadromous fish, such as brook trout and lake whitefish, entered the waters of James Bay to spawn from rivers that would have been disrupted or destroyed by additional extensive hydroelectric development under James Bay II. According to the Sierra Club (n.d.), "Estuaries, heath-covered islands, salt marshes, freshwater fens, sub-tidal eelgrass beds, and ribbon bogs in the James Bay region provide nourishment for huge flocks of geese, ducks, and loons. Tundra on either side of the bays provides habitat for caribou, moose, otter, muskrat, beaver, lynx, and polar bear. This wildlife, in turn, supports the traditional Cree trappers and fishers of the James Bay region and, along Hudson Bay, the Inuit and Naskapi. The Cree regard their part of the eco-region as a 'garden' providing for all their needs."

The Cree and Inuit have adapted to this world of forest, flowing water, and marsh since the end of the last ice age about 10,000 years ago. At the end of the twentieth century on Europe's clock, Hydro-Quebec proposed to alter the entire ecosystem and destroy large parts of it to provide electricity to urban areas in more temperate reaches. Commented one observer,

> Hydro projects could have far-reaching and negative impacts if, for example, breeding waterfowl habitat in Quebec river estuaries was damaged, thereby affecting migratory bird populations wintering in the eastern and southern United States. ... Approximately 75 percent of the global population of Atlantic brant geese are concentrated on the eel grass beds of the Quebec coast and parts of the Ontario coast of James Bay, and almost the entire North American population (up to 320,000) of black scoters use southern James Bay as a staging area. Other waterfowl species that utilize inshore, inter-tidal and brackish coastal habitats in the Hudson Bay/James Bay bio-region include black duck, pintail, mallard, wigeon, green-winged teal, and scaup. Mergansers and loons make extensive use of offshore water for feeding, and a significant number of common eider pass the winter in James Bay and the Belcher Islands. ("Sustainable Development" n.d.)

For several years during the early and mid-1990s, Cree leaders and elders utilized the lecture circuit with the story of how Hydro-Quebec's plans would damage their world. Bill Namagoose, a James Bay Cree, was quoted in July 1993 as saying,

> We're up against the perception that Hydro-Quebec is the engine [of progress] and they've used it to whip up the nationalism of Quebecers against the Cree and the Inuit. [However], if you cut off the market there's no point in doing the financing. If there is no financing there is no construction. And so far we have managed to cancel two American contracts. There is no economic justification for the project, no environmental, no energy reason why these projects should be built. We've been living up there in harmony with the environment. But when the invasion began in 1970, we were totally overwhelmed. . . . We've survived as a nation for many thousands of years and we [would] like to arrive in the future on our own terms and traditions . . . living on an environmentally sound land, with all of our rivers intact for thousands and thousands of years. Two-fifths of our land has already been impacted by the first phase. We know what it's like to live in the middle of a mega-project. That's why we're trying so hard to protect the parts of our lands we still have left. (Vance and Shafer 1995)

Albert Meascum, a Cree trapper living in the village of Ouje-Bougoumou, said that he had asked the company not to cut certain parts of his trapping territory, but they paid no attention. "Before the forest companies came," he said, "there was lots of game. By cutting down the trees they have chased away the animals." The companies also pollute, he said, as they spill oil and gas on the land and poison the young willows that feed the moose. "When the snow melts," he said, "the gas and oil go into the lakes and pollute the water and the fish" (Native Forest Network 1994).

The future of the James Bay project was litigated for several years. During September 1991, a Canadian federal judge in Ottawa ordered a new assessment of the James Bay II project under a process that could give federal authorities the right to stop it. "I conclude that the Crees' right to an independent parallel federal review must be honored," wrote Justice Paul Rouleau ("Cree Indians" 1991, 4). At the time, the total anticipated cost of James Bay I and II was estimated to be $62 billion.

In February 1994, the Supreme Court of Canada ruled that the National Energy Board had the right to examine the environmental effects of the Great Whale Project, essentially the same issue that had been decided by Judge Rouleau two and a half years earlier. Armand Couture, president of Hydro-Quebec, denied at the time that the court's order would have any impact on plans to throw the switch on Great Whale II by 2003.

In the meantime, the Crees continued to stoke popular pressure against Hydro-Quebec's efforts to market power from the James Bay projects. Ten Cree and Inuit activists paddled a 25-foot combination kayak/canoe from Ottawa to New York City during the spring of 1990. They crossed the Quebec–Vermont border at the northern end of Lake Champlain and timed their arrival in Central Park to coincide with Earth Day. During the first week of October 1991, Cornell

University's American Indian Program organized a forum on James Bay development, with Mathew Coon-Come as keynote speaker. In mid-April 1993, Native leaders and their supporters bought a full-page advertisement opposing James Bay II in the *New York Times*. Robert Kennedy Jr. and Coon-Come held a joint press conference, also in New York City, yet another high-profile action that turned more New Yorkers against the project. New York State then yielded to growing consumer protests and canceled a $17 billion export contract with Hydro-Quebec.

On November 18, 1994 (a few months after Hydro-Quebec's president, Armand Couture, had asserted that electricity would begin to flow from James Bay II in 2003), Quebec Premier Jacques Parizeau said that the project was being shelved "indefinitely." He told the press, "We've not saying never, but that project is on ice for quite a while." The decision was made after two governmental review committees said that Hydro-Quebec "would have to go back to the drawing board to correct 'major inadequacies' in its environmental assessment of the project." Hydro-Quebec had been stopped by a blizzard of paper; company officials complained that its environmental impact study, which ran to 5,000 pages and took 11 years to compile, had cost $190 million ("Quebec Will Shelve" 1994, 7).

## GREAT WHALE REDUX

The Great Whale (James Bay II) hydroelectric project was revived on a reduced scale during 2001 after having been shelved seven years earlier. The project was revived with a new twist: a proposal that the Crees, who had played a large role in opposing the project, would eventually become its sole owners. The idea was proposed to the Crees by Toronto-based AMEC Inc., a large engineering and construction firm that had worked on Hydro-Quebec's original James Bay hydroelectric project as well as China's Three Gorges Dam and Colombia's Urra Dam. The company's name was Agra Inc. until it was bought in 2001 by AMEC PLC, a British-based conglomerate with 50,000 employees in 40 countries.

The band councils of Whapmagoostui and neighboring Chisasibi held quiet discussions addressing the idea of reviving Great Whale in 2001. "It's just an idea, it's not official," said David Masty, a chief at Whapmagoostui. "It's at the discussion stage. We haven't taken a position on it as a community." He continued, "Some Cree officials are furious that the talks have gone on with virtually no community involvement. They say the discussions are a slap in the face to those who fought to kill the earlier variants of the project. They also point out that Crees have repeatedly voted—in regional assemblies, band-council votes and a community referendum—against new hydro projects" (Roslin 2000).

The new proposal involved diversion of the Great Whale River at its headwater, Lac Bienville. At an estimated cost of $350 million, the river's water would be redirected through 10 kilometers of canals into Hydro-Quebec's La Grande hydroelectric complex, increasing the water flow through the turbines. In 1997, Hydro-Quebec had proposed to divert the Great Whale and, for the first time, sought the Crees' consent and offered a minority partnership in the project.

This proposal was shelved after 92 percent of Whapmagoostui Crees voted against any development projects on the river (Roslin 2000). The new proposal retained the diversion idea but proposed that the Cree would own the facility once its debt was paid off. AMEC would design the facility, arrange financing, and retain an ownership share while the debt was being paid off.

Hydro-Quebec, which earned a net profit of more than $1 billion in 2000, wanted to construct the dams to increase its profits by selling surplus power outside Quebec, which already produced enough power to meet its own needs. Hydro-Quebec planned to allow small private energy producers to build and operate the dams. In turn, the owners of the smaller dams would sell their electricity to Hydro-Quebec. According to an account in the *Plain Dealer*, "Quebec has turned its back on the environment, suggesting that even the Rouge River in the beautiful Laurentians could be a site for a dam. Thousands visit the area every year to paddle the white water rapids and pristine pools" (Egan 2001, D-15).

The Quebec government and natives in the James Bay region signed an agreement in principle on October 23, 2001, to end a bitter legal dispute over hydroelectric development in the area and open the door to further resource development in northern Quebec. Under the agreement, the Cree will receive Canadian $3.5 billion (roughly US $2.2 billion at the time) worth of natural resources royalties from the province over 50 years in return for giving up all legal action against a major power project in the region. "This agreement constitutes, I am sure, the basis of a great peace between Quebec and the Crees," Quebec Premier Bernard Landry said before a signing ceremony with Cree Grand Chief Ted Moses. "This is an historic turning point and a truly profound revolutionary vision for the Cree and aboriginal peoples generally," Moses said (Conférence 2001). The project, which had been on hold for several years because of opposition from the Cree, should produce 1,300 megawatts, or about 15 percent of the power from the entire James Bay area, and create about 8,000 jobs during six years of construction.

The agreement, which was later ratified by the Cree communities of northern Quebec, allowed the 15,000 natives an annual revenue flow and direct participation in any economic development on native land in northern Quebec. The agreement pledged to pay the Cree $70 million a year for 50 years. It also included Hydro-Quebec jobs for Crees, an important issue in communities where more than 80 percent of the young people under 25 years of age are unemployed during the winter. The agreement also promises remediation of mercury contamination, funding for start-up business programs, job training, health and social services, electricity, sanitation, and fire services for Cree communities (Matteo 2002).

In return, the Crees promised access to resources (including diversion of the Rupert River for hydroelectric development) and cession of $3.6 billion in environmental lawsuits. The proposed diversions of the Rupert River are subject to environmental impact reviews, but the government of Quebec is allowed to make the final decisions. "In effect," according to one analysis, "the Cree will not be able to protest, stop, inhibit, or litigate" environmental outcomes, including the anticipated drying up of parts of the Eastman River and the flooding of traplines (Matteo 2002).

The Cree agreed to drop their environmental lawsuits in part because legal fees were costing them $9 million a year but also because many of the cases were faring badly in the Canadian court system. Courts in Quebec had ruled against the Cree on forestry issues as well as their opposition to hydroelectric development on the Eastman River (Matteo 2002). The agreement also is important because it saves roughly 8,000 square kilometers of land from being flooded, according to Coon-Come. "We want jobs," he said after the new agreement was negotiated. "We want a say in where development takes place [and] what happens in our own backyard" (Matteo 2002). "It isn't a blank check," said chief Cree negotiator Abel Bosum. "Quebec is taking a big risk here. . . . We consent to Eastmain-Rupert but there is no guarantee that it will be approved." He explained that the project must pass impact studies by Quebec, Ottawa, and the Cree. "It's not a done deal," he said (Dougherty 2001).

Cree spokesman Romeo Saganash recalled that in 1975, the original James Bay agreement also was greeted with hope. Instead, the Cree faced 26 years of frustration, fighting Quebec and Ottawa in the courts. "Today my only hope is that my children will not be in front of your children in 25 years' time saying: 'Here's another agreement. After 25 years of frustration, we couldn't implement the deal we signed in 2001' " (Dougherty 2001).

Once the Crees' tentative agreement with Hydro-Quebec was negotiated, an intense debate began among the Cree as to whether it should be accepted, with two Cree leaders taking opposite sides. On December 18, 2001, Grand Chief Moses denied that the tentative deal was unraveling. "Sure, there is criticism and suspicion," Moses told the Quebec Electricity Association. "Why should this be surprising? And why in any society should this democratic debate be interpreted as a sign of failure? We have been in a long and difficult struggle to achieve recognition and respect for the Cree people." Moses said that opponents of the agreement were "a small group of people trying to make a lot of noise" (Macafee 2001).

"I stand firm in my opposition to the deal," Matthew Mukash, deputy grand chief, said shortly after arriving in Montreal for a Cree Grand Council meeting. "The Cree Nation is clearly divided on this. It's difficult at this time to know which way they're going to go." Mukash said that he is concerned that aboriginal rights could eventually be extinguished under the deal. There is also an implication the agreement would reaffirm Quebec's position that it owns the resources on the land, not the Cree. "I think it's very dangerous when you look at it over the long term," said Mukash (Macafee 2001).

During late January 2002, Quebec Crees voted in favor of the $3.4 billion deal with the provincial government. Of the 6,500 eligible voters, 69.35 percent, or 3,106, voted in favor of the deal, with 30.65 percent, or 1,373, against. The decision, which Grand Chief Moses called historic, meant that the Cree nation was ready to complete a process that it had started in October when it signed an agreement in principle with Premier Landry to allow hydroinstallations along the

Eastmain and Rupert rivers subject to environmental approval. One Cree leader, George Wapachee, said that the Cree who oppose hydroelectric development have sat on the fence too long while the world passes them by. In his community, the population has jumped to 500 and will hit 1,000 in 20 years. The land can support only so much traditional trapping and logging activity, he said. "There's a lot of young people who have nothing to do," said Wapachee. "We have to look forward, move ahead. That's the way life is" (Authier 2002).

As Grand Chief Moses and Premier Landry signed the hydroelectric power deal in the Cree village of Waskaganish on February 7, 2002, an elderly Cree chief, Henry Diamond, broke into the assembly hall, shouting his objections. He was tackled by seven police officers and hauled away, bleeding from the head, to be charged with disturbing the peace and resisting arrest. Another dissident, Chisasibi Band Councilor Larry House, tried to voice his objections at a press conference following the signing. He was arrested for shoving an officer and impeding pedestrian traffic, for which he was fined $25 (Taylor 2002). Opposition to the agreement was centered in the Chisasibi community, which sits at the foot of a reservoir that has been plagued with technical problems. The reservoir also will rise at least six feet once the Rupert River is diverted, causing floods.

"Today," said Cree Deputy Chief Matthew Mukash, "Native communities across Canada are calling us sellouts." Mukash believes that the deal is a major mistake that will eventually lead to "the gradual and progressive takeover of the Cree by Canada and Quebec." He pointed out that while 72 percent of the Cree who voted supported the agreement, only 53 percent of those eligible to vote did so. Many who did not vote "no" did so from fear, Mukash said (Matteo 2002).

## *THE PIMICIKAMAK CREE OF MANITOBA: IMPERILED BY HYDROPOWER*

The Pimicikamak Cree assert that, for 25 years, the Manitoba Hydro Power Company has been destroying their traditional lands and society with a series of dams and canals cut through wilderness to divert the flow of rivers on which their traditional economy depends. The Crees signed a treaty in 1977 that said they would receive full compensation for damage to their homeland. They assert, however, that this treaty has not been honored by Manitoba Hydro. Now the Pimicikamak Crees, who have been experiencing 70 to 80 percent unemployment, have lost their traditional economic base and their cultural and spiritual foundations, leading to an epidemic of suicides (Peterson 2000).

Dams built during the 1970s have reengineered Manitoba's two largest rivers and largest lake to serve the energy needs of Manitobans and United States export customers. For 25 years, unnaturally fluctuating waters in Pimicikamak Cree territory have eroded shorelines. According to advocates for the Cree, "Burial grounds have been exposed, islands eroded right off the map, [the] ecological balance [has been] disrupted, and large areas of Cree fishing and trapping grounds have become despoiled and inaccessible." Cross Lake, immediately downstream of the Jenpeg Dam, has lost as much as 300 square kilometers, about half of its surface area, because the water has been held in Lake Winnipeg to be released during times of peak electrical demand in winter. Although Manitoba Hydro constructed a rock weir in 1991 (14 years after committing to build it), "navigation remains treacherous and receding waters still expose muddy flats at certain times" ("Hydroelectric" 2001).

The debate over James Bay II continued in tandem with an environmental analysis of seven similar dams that were built during the early 1970s in the Nelson and Churchill river systems of Manitoba. These rivers drain into the west coast of Hudson Bay, of which James Bay is a part. By the early 1990s, studies showed that one in six people in the area was suffering from mercury contamination (LaDuke 1993, A-3). A large percentage of game had disappeared from the affected area, making subsistence hunting very difficult. Several hundred Native people were displaced into housing projects as their homes were flooded.

The environmental impact assessment for the Churchill–Nelson project was an after-the-fact matter; it was initiated after Manitoba Hydro had fixed the configuration, operating regime, and timing of construction for the diversion of the Churchill River into the Nelson River. The unforeseen environmental impacts of the Manitoba projects were described as follows:

> Severe shoreline erosion, caused by impoundment of Southern Indian Lake as part of the diversion, led to increased turbidity. This resulted in the collapse of the commercial whitefish industry. Whitefish populations in Cross Lake, another large lake in the Churchill-Nelson river basin, fell 65 percent. In addition, walleye and northern pike in all flooded lakes along the diversion route accumulated mercury levels that exceeded Canadian limits for the protection of human health. Some commercial fisheries were permanently closed, and local residents were encouraged to avoid consumption. The collapse of the commercial fishery placed a severe strain on the social fabric of northern aboriginal communities, forcing them to move and/or to rely increasingly on compensation payments for income. ("Sustainable Development" n.d.)

Methylmercury contamination was detected in northern Manitoba during the early 1970s by scientists from the Freshwater Institute in Winnipeg. Similar contamination had not been anticipated by Hydro-Quebec as it began construction of James Bay Phase I at the end of the decade despite the experiences of a similar utility, doing similar construction work, geographically adjacent. Apparently, no one was paying attention.

The Pimicikamak Cree held a "Hands Across the Border" conference in 2001 to strengthen its campaign to build a coalition of native nations, farmers, consumers, and environmentalists in the United States and Canada to force the cancellation of utility contracts with Manitoba Hydro requiring a new megadam project. Five existing dams in Manitoba have submerged 3.3 million acres of land, including large tracts of boreal forest and extensive animal habitat. The methylmercury created by the construction of these dams has contaminated fish, a staple of the Cree diet, in the Nelson River.

The Pimicikamak Crees of Manitoba adopted a strategy similar to that of the Quebec Cree, who forced Hydro-Quebec to shelve its James Bay Phase II project in the mid-1990s. They traveled to the United States to convince consumers of the utilities that may purchase the electricity generated by the dams to advocate for cancellation of their contracts on moral grounds. As part of this effort, the Pimicikamak Cree lobbied Minnesota public utilities to weigh damage to the native peoples' ecosystems when purchasing power from Manitoba.

In late 2000, the Minnesota Public Utilities Commission agreed to include ecological damage in its decision making, a decision that Andy Orkin, one of the Crees' attorneys, called "a big win for the environment" ("Cree Claim" 2000). The decision required the commission to assess how energy projects would affect the environment and the people who live near them. For years, electricity generated in Manitoba had been sold in Minnesota, but the state's Public Utilities Commission never had investigated the human impact of power generation. This decision was intended to create a methodology that will compare coal, nuclear, hydroelectric, wind, and other generation methods with each other in terms of socioeconomic and environmental impacts.

Orkin said the commission's decision meant that the impact of Manitoba Hydro's hydroelectricity projects on native peoples will come under closer scrutiny. The renewal of Manitoba Hydro's contract with Minnesota utilities was upheld, however. "What the Pimicikamak Cree Nation was much more concerned about was protecting the Lake Winnipeg, Churchill, Nelson Rivers' environment, particularly in and around [the] Pimicikamak Cree Nation's traditional territory, against the impacts of further large-scale hydroelectric development by Manitoba Hydro," said Orkin ("Cree Claim" 2000).

Pimicikamak Cree Nation Chief John Miswagon traveled to Minneapolis to attend an annual shareholder's meeting of Xcel Energy, which purchases approximately 40 percent of Manitoba Hydro's power output, supplying power to 12 states south of the border. Miswagon addressed 1,800 shareholders and utility managers as they voted on a resolution demanding that the company withdraw from Manitoba Hydro and increase its power supply from renewable resources that do not negatively impact human rights and the environment. The motion was organized by the As You Sow Foundation, a U.S. organization that advocates corporate responsibility. The resolution was rejected, but 8.8 percent of shareholders voted in favor of it. A 3 percent vote was required to place the resolution on the Xcel's Corporation's agenda for the following year ("Cross Lake" 2001). To promote its cause, the Crees bought a full-page advertisement in the *New York Times*, purchased for about $100,000, with funding from an anonymous donor.

Michael Passoff, a corporate responsibility expert with As You Sow, said the vote indicated a high level of support for change among Xcel shareholders. "The vote . . . sends a clear message to management that this issue is likely to grow unless the company addresses shareholder concerns about the destructive environmental and human rights impact of increasing energy imports to the U.S. from

Manitoba Hydro's mega-projects" ("Cross Lake" 2001). Miswagon said the vote was clearly a success for the first nation. "This is a major victory . . . in our struggle for the environment and human rights," Miswagon said. "We are blown away by this level of compassion and support from American shareholders" ("Cross Lake" 2001).

Manitoba Hydro maintained that it had spent $400 million (Canadian) over 20 years to compensate northern communities, but there appeared to be little to show for it. The Crees and other native peoples asserted that poverty, unemployment, and suicide among the Pimicikamak Cree may be attributed to hydroelectric development ("Cree Claim" 2000). George M. Ross, a Cree elder, said, "The south is benefiting at the expense of our misery" ("Hydroelectric" 2001).

## *RESISTANCE TO A HYDROELECTRIC PROJECT IN LABRADOR*

Inuit in Labrador have been resisting hydroelectric development in a fashion similar to peoples in Quebec and Manitoba. In this case, they oppose the Churchill River's hydroelectric dam. This struggle occurs within a broader context of a movement to gain official recognition as Inuit descendants. Unlike their kin in Nunavut to their north and west (who have a semisovereign state), the Labrador Inuit are officially invisible. For many years, the same people (known as Inuit-Métis or Labrador Métis) have been engaged in a struggle for official recognition as part of a group descended from signers of a treaty with Europeans in 1765. Today, they are united in the NunatuKavut (meaning "our ancient land") Community Council.

"We've been pushed around for generations," said Todd Russell, president of the NunatuKavut Community Council (formerly the Labrador Métis Association), who was arrested with Elder James Learning as they blocked roads in protest of the planned $7.7 billion Muskrat Falls project, which involves a dam, a power station, and a web of huge transmission lines to Newfoundland and Nova Scotia. Their blockade of the Trans-Labrador Highway on April 5, 2013, was in defiance of a judge's permanent injunction against disruption of construction, publicizing their exclusion from negotiations regarding the power complex.

"We will defend ourselves in the court system, but we will continue to assert our aboriginal rights to our traditional territory, and we will continue to mount protest after protest if that's what it takes to have our views known and our rights respected. . . . It's the area where we hunt, where we fish, where we have built homes, where our people have trapped," said Russell, a former liberal member of Parliament. "There are areas of a sacred, and very special, nature there. The government will not recognize that there are overlapping and conflicting interests with this hydroelectric development" (Ball 2013).

During his arrest, after he had lay down on the highway with other protesters, Russell "was dragged by members of the Royal Canadian Mounted Police (RCMP) after he lay down alongside the other protesters, all arrested on obstruction charges. Though seven were released on bail the same day, Learning refused to sign a written promise to stay off the land on the grounds that doing so would extinguish his aboriginal title and rights to his people's traditional territories" (Ball 2013). He remained imprisoned in the Labrador Correctional Centre at Goose Bay and went on a hunger strike despite the fact that he was ailing from prostate cancer that was spreading to his bones.

"It is tragic that our father has had to risk death through hunger to protest the destruction of his homeland and culture, of NCC [NunatuKavut Community Council] territory and culture," said Learning's daughter, Carren Dujela, in a statement before his release. "How do you tell your children their grandfather is in jail and on a hunger strike? With tears in your eyes and pride in your heart!" (Ball 2013).

## FURTHER READING


Authier, Philip. "Cree Pact Clears Final Hurdle." *Montreal Gazette*, February 5, 2002. http://www.canada.com/montreal/montrealgazette/story.asp?id={43F9B38E-7577-4579-AEED-90A8F9638936}

Ball, David P. 2013. "Enbridge Faces Rising Opposition to Northern Gateway Pipeline with Protests, Arrests." Indian Country Today Media Network, January 25. http://indiancountrytodaymedianetwork.com/2013/04/04/red-lake-band-chippewa-determined-blockade-atop-four-enbridge-pipelines-minnesota-148600

Barreiro, Jose, ed. 1991. "Readings on James Bay." *Akwe:kon Journal* 8, no. 4 (Winter): n.p.

Bigert, Claus. 1995. "A People Called Empty." In *Amazon of the North: James Bay Revisited* by Rainer Wittenborn and Claus Biegert. Program for show, Santa Fe Center for Contemporary Arts, August 4–September 5, n.p.

"Canadian Indians Paddle to New York City to Protest Quebec Power Plant." 1990. *Syracuse Post-Standard*, April 5, A-2.

"Conférence de presse de M. Bernard Landry, premier ministre du Québec, et de M. Ted Moses, grand chef du Grand Conseil des cris Le mardi 23 octobre 2001." 2001. Government of Quebec. October 23. http://www.assnat.qc.ca/fra/conf-presse/2001/011023BL.htm

Coon-Come, Matthew. 2001. "Remarks of the National Chief Matthew Coon-Come." Peoples Summit of the Americas Environment Forum, April 18. http://www.afn.ca/Press%20Realeses%20%20speeches/april_18.htm

"Cree Claim Victory in Hydro Hearings." 2000. Canadian Broadcasting Corporation, December 1. http://www.northstar.sierraclub.org/HYDRO_Cree_Claim%20_%20Victory.htm

"Cree Indians Win Battle in Their Struggle to Stop Canadian Hydropower Plant." 1991. *Omaha World-Herald*, September 12, 4.

"Cross Lake Takes Hydro Battle to U.S." 2001. First Perspectives: News of Indigenous Peoples of Canada, May 29. http://collection.nlc-bnc.ca/100/201/300/first_perspective/2001/05-29/story7.html

Dougherty, Kevin. 2001. "Deal Is No Blank Cheque, Crees Say." *Montreal Gazette*, October 24. http://www.canada.com/montreal/montrealgazette/story.asp?id={4FCA2E44-DDE4-4675-88C5-DEFD94486151}

Dumont, C. 1995. *Proceedings of 1995 Canadian Mercury Network Workshop. Mercury and Health: The James Bay Cree*. Montreal: Cree Board of Health and Social Services. http://www.cciw.ca/eman-temp/reports/publications/mercury95/part4.html

Dumont, Charles, Manon Girard, François Bellavance, and Francine Noël. 1998. "Mercury Levels in the Cree Population of James Bay, Quebec, from 1988 to 1993/94." *Canadian Medical Association Journal* 158 (June 2): 1439–45. http://www.cma.ca/cmaj/vol-158/issue-11/1439.htm

Egan, D'Arcy. 2001. "Dam the Ecology If Quebec Goes Full-Speed Ahead." *Plain Dealer*, July 29, D-15.

"Hydroelectric Production Is Anything but Cheap." 2001. *Virtual Circle: First Nations' Chronicles*, March 21. http://www.vcircle.com/journal/archive/010.shtml

LaDuke, Winona. 1993. "Tribal Coalition Dams Hydro-Quebec Project." *Indian Country Today*, July 21, A-3.

Macafee, Michelle. 2001. "Cree Grand Chief Defends Hydro Deal with Quebec as Negotiations Wind Down." *Ottawa Citizen*, December 18. http://www.canada.com/news/story.asp?id={565BD3CB-8DAB-4CD2-878F-EE2F7807666B}

Matteo, Enzo di. 2002. "Damned Deal: Cree Leaders Call Hydro Pact Signed in Secret a Monstrous Sellout." *Now Magazine*, February. http://www.nowtoronto.com/issues/2002-02-14/news_story.php

McKeown-Eyssen G. E., and J. Ruedy. 1983. "Methyl-Mercury Exposure in Northern Quebec: Neurologic Findings in Adults." *American Journal of Epidemiology* 118: 461–69.

Native Forest Network. 1994. "Quebec-Hydro Project May Destroy James Bay." Friends of the Earth, Victoria, British Columbia. http://jinx.sistm.unsw.edu.au/~greenlft/1994/137/137p14.htm

Peterson, Diane J. 2000. "Two Twin Cities Churches Protest Hydro Power Injustices." *Earthkeeping News* 9, no. 5 (July/August). http://www.nacce.org/2000/manitoba.html

"Quebec Will Shelve Huge Hydroelectric Project Indefinitely." 1994. *Omaha World-Herald*, November 19, 7.

Roslin, Alex. 2000. "Crees Revive Hydro Project." *Montreal Gazette*, January 21. http://www.montrealgazette.com/news/pages/010121/5036705.html

Sierra Club. N.d. "Hudson Bay/James Bay Watershed Eco-Region." http://www.sierraclub.org/ecoregions/hudsonbay.asp

"Sustainable Development in the Hudson Bay: James Bay Bio-Region Canadian Arctic Resources Committee." N.d. Environmental Committee of Sanikiluaq, Rawson Academy of Aquatic Science. http://www.carc.org/pubs/v19no3/2.htm

Taylor, Robert. 2002. "Cree Leaders Jailed in Protest over Power Deal." *Indian Country Today*, February 17. http://www.indiancountry.com/?1013955409

Vance, Chris, and John Shafer. 1995. "Voices of Resistance." Radio transcript, CFUV 102 FM (community radio station, University of Victoria, British Columbia), transcribed in 1995 and reprinted in the magazine *All That's Left Is Struggle*. http://www.finearts.uvic.ca/~vipirg/SISIS/sov/allcree.html

# Chapter 4

## *Climate Change and Native Peoples*

Climate change induced by human consumption of fossil fuels impacts everyone on Earth and has become the signature environmental issue of our time. Native peoples of North America, with their close connection to the Earth and subsistence lifestyles, are among the first to be the most intensely affected by a rapidly changing climate. This is most evident in the Arctic, which is the most rapidly warming area on Earth and where an Inuit world adapted to life on ice is melting away. Alaskan Native communities also face climate-induced change, including the relocation of entire coastal villages. Elsewhere in North America, Native water resources and food sources already have been significantly damaged by a warming climate.

"We need to be worried about climate change because it's clearly already affecting our region in ways that impact many areas—we're seeing landscapes burning, dying because of heat and dryness," Jonathan Overpeck, of the Southwest Climate Science Center, told the Indian Country Today Media Network. "We're seeing reservoirs that were full just ten years ago now only half full on the Colorado [River]. These are visible harbingers of what might come. What we need to do as a society is talk about it and figure out how to deal with these challenges" (Allen 2012).

### *AMERICAN INDIANS FEEL THE EFFECTS OF CLIMATE CHANGE MOST INTENSELY*

A National Wildlife Federation report, *Facing the Storm: Indian Tribes, Climate-Induced Weather Extremes and the Future for Indian Country*, suggests that American Indians suffer more intensely than other ethnic groups from climate change. The report said that "the high dependence of tribes upon their lands and natural resources to sustain their economic, cultural and spiritual practices, the relatively poor state of their infrastructure and the great need for financial and technical resources to recover from such events all contribute to the disproportionate impact on tribes." Not only does Native American spirituality emphasize relationships with nature, but Alaska Natives and American Indians depend more on natural resources, as well as rivers and oceans, the report said: "Extreme weather events can be very destructive for tribes, many of whom are already suffering from lack of resources to begin with," said Amanda Staudt, a scientist at the National Wildlife Federation and a contributor to the report ("American Indians" 2011).

"Climate change is reshaping the daily lives of many Native peoples," according to reporting on the Indian Country Today Media Network, "especially that of coastal communities. They have witnessed the destruction of their freshwater streams, lakes, watersheds, coral reefs, and the devastating decline of fish and wildlife." At a First Stewards Symposium convened during July 2012 in Washington, D.C., coastal tribes and nations of Washington State (Makah, Quileute, Hoh, and Quinault Indian) drafted a resolution sent to the U.S. Congress and President Obama that "requests formal recognition of the coastal indigenous people and their expertise in understanding and adapting to changes in their natural systems." "We are girding our loincloths and sharpening our spears to undertake this project and are asking President Obama, the U.S. Congress and others to engage in sincere and earnest consultation with us, so our cultures, our peoples and our world can survive and thrive," said Kitty Simonds, vice chairman of the First Stewards board of directors ("First Stewards" 2012).

## *CONSULTING WITH NATIVE PEOPLES*

Greenpeace activists have been gathering information about global warming from Yup'ik and Inupiat Inuit. Greenpeace and the Arctic Network, a nonprofit conservation group in Alaska, also have been interviewing indigenous people of the Arctic about the condition of the ice with which they live on a daily basis. These anecdotal accounts support statistics indicating that Alaska's climate is warming more rapidly than perhaps any other place on Earth due to increased proportions of carbon dioxide and other greenhouse gases in the atmosphere.

The activists reported that Savoonga whalers' harvests have been declining due to the early breakup of ice, forcing them to return home earlier than usual. Inuit said that for several years, they also have been routinely catching fish heretofore not considered native to the Arctic, including chum, pink, and silver salmon. Like observers all across of the Arctic, Inuit interviewed by Greenpeace said they had observed birds that are new to higher latitudes, such as various types of swallows, finches, and robins.

Most of global warming's debits are accruing to the Inuit, the animals of the region, and the traditional relationship between the two. Precipitation may increase as much as 25 percent with the advent of rapidly rising temperatures during the twenty-first century. When the precipitation falls as snow and ice, caribou and other grazing animals will have a harder time finding food during lean winter months. The study anticipates that heavier snow cover will lead to smaller, thinner animals that will be forced to go farther afield for food in winter. The same animals will be plagued by increasing numbers of insects during warmer summers. Environmental stresses during the next century could lead to "complete reproductive failure of the caribou in a worst-case scenario" (George 2000).

The U.S. Geological Survey (USGS) also has been consulting with Arctic peoples. "Many climate change studies are conducted on a large scale, and there is a great deal of uncertainty regarding how climate change will impact specific regions," said USGS social scientist Nicole Herman-Mercer. The USGS, for example, described thinning ice on the Andreafsky and Yukon rivers, as it noted that Arctic peoples use frozen rivers as transportation routes on snowmobiles or dogsleds. Early melting and late freezing disrupt hunting patterns and trade.

Alaska Natives (such as Yup'ik in St. Mary's and Pitka's Point) have told scientists that weather has become more unstable in ways that affect their lives. For example, less melting snow means that thawing rivers carry less of the driftwood that Native peoples use for cooking and heat. In Point Hope, Alaska, 330 miles south of Point Barrow, a 4,500-year-old Alaskan native village of 900 people, wrote Juliet Eilperin (2012) in the *Washington Post*,

> fermented whale's tail doesn't taste the same when the ice cellars flood. . . . Whaling crews in this Arctic coast village store six feet of tail—skin, blubber and bone—underground from spring until fall. The tail freezes slowly while fermenting and taking on the flavor of the earth. Paying homage to their connection to the frozen sea, villagers eat the delicacy to celebrate the moment when the Arctic's ice touches shore. But climate change, with its more intense storms, melting permafrost and soil erosion, is causing the ice cellars to disintegrate. Many have washed out to sea in recent decades. The remaining ones regularly flood in the spring, which can spoil the meat and blubber, and release scents that attract polar bears.

Stored in an electric freezer, the fermented whale's tail loses its distinctive taste. "So much of our culture is being washed away in the ocean," said Point Hope Mayor Steve Oomittuk. "We live this cycle of life, which we know because it's been passed from generation to generation. We see that cycle breaking" (Eilperin 2012). Melting ice deprives hunters of firm footing and obliterates paths that once were reliable. Instead of poking their heads through holes in the ice, seals remain submerged in water, evading hunters.

The cascading effects of climate change on North American Native peoples became the sole focus of an entire issue of the scientific Journal *Climatic Change* published in October 2013. Special attention was paid to the congruence between Native elders' observations and meteorological records, finding widespread agreement. Native subsistence was a major focus of this scientific focus, ranging from observations that "hunters have observed that moose seem less healthy" (Lynn et al. 2013, 549) to effects of diminishing sea ice on walrus, effects on salmon runs, and damage from increased storm intensity and rising seas. Some shore-dwelling tribes along the Washington coast (such as the Quileute and Hoh) have moved or are considering leaving seafront homelands inundated by rising storm surges.

Many Native residents of the Arctic have asserted that the weather with which they live has become stormier as the weather has warmed. In late 2009, two scientists in the School of Earth Sciences at the University of Melbourne, Victoria, Australia, provided statistics that support those assertions. Ian Simmonds and Kevin Keay studied the month of September because sea ice has been markedly reduced from 1979 through 2008 and found that cyclonic storms had become stronger, although the number of storms had not changed significantly. "The findings reinforce suggestions that the decline in the extent and thickness of Arctic ice has started to render it particularly vulnerable to future anomalous cyclonic activity and atmospheric forcing," they wrote in *Geophysical Research Letters* (Simmonds and Keay 2009).

## *AN ICE WORLD MELTS*

Ice is fundamental to the entire Arctic food chain:

> Sea ice is the foundation of the Arctic marine environment. Vital organisms live underneath and within the ice itself, which is not solid, but pierced with channels and tunnels large, small, and smaller. Trillions of diatoms, zooplankton, and crustaceans pepper the ice column. In spring, sunlight penetrates the ice, triggering algal blooms. The algae sink to the bottom, and in shallow continental shelf areas they sustain a food web that include clams, sea stars, arctic cod, seals, walruses,—and polar bears. (McGrath 2011, 70)

The swiftly changing Inuit world revolves around the ice, which is melting rapidly year by year. The traditional economy and its spiritual bonds are tethered to the ice. The pronounced thinning of Arctic sea ice has made the ice pack more brittle and susceptible to wind drift. The volume of Arctic sea ice decreased by one-third during 2007–2011 compared with the 1979–2006 mean. As Richard Kerr (2012, 1591) wrote in *Science*,

> The now-clearly-accelerating decline of summer ice—punctuated by exceptional losses in 2007 and now in 2012—has persuaded everyone that summer Arctic sea ice will be a goner far sooner than the end of the century, as current models predict. So the full knock-on effects of an ice-free Arctic Ocean—from the loss of polar bear habitat to possible increases of weather extremes at midlatitudes—could be here in many people's lifetimes. How far wrong the models might be, however, is still very much in dispute.

"It's hard even for people like me to believe, to see that climate change is actually doing what our worst fears dictated," said Jennifer A. Francis, a Rutgers University scientist who studies the effect of sea ice on weather patterns. "It's starting to give me chills, to tell you the truth" (Gillis 2012). The low of 3.41 million square miles of water-laced ice in the summer of 2012 was 18 percent below the 2007 record (an area as large as Texas) and half of the area measured by the first satellite photos in 1979. Between 1979 and 2001, ice declined in the Arctic on an average of 6.5 percent per year. In the next 11 years, the average rate of decline doubled. "The rules are changing," said Mark Serreze of the National Snow and Ice Data Center. "The ice is so thin, it just can't survive the summer" (Kerr 2012, 1591).

"When the ice thins to a vulnerable state, the bottom will drop out and we may quickly move into a new, seasonally ice-free state of the Arctic," Serreze said. "I think there is some evidence that we may have reached that tipping point, and the impacts will not be confined to the Arctic region" ("As Arctic Sea Ice Melts" 2008). The review paper by Serreze and Julienne Stroeve of the National Snow and Ice Data Center at the University of Colorado, Boulder, and Dr. Marika Holland of the National Center for Atmospheric Research titled "Perspectives on the Arctic's Shrinking Sea Ice Cover" appeared in the March 16, 2007, issue of *Science*. "We're seeing more melting of multi-year ice in the summer," said Stroeve. "We may soon reach a threshold beyond which the sea ice can no longer recover." "We have already witnessed major losses in sea ice, but our research suggests that the decrease over the next few decades could be far more dramatic than anything that has happened so far," said Holland. "These changes are surprisingly rapid" ("As Arctic Sea Ice Melts" 2008).

In August and September 2012, sea ice covered less of the Arctic Ocean than at any other time since at least 1979, when the first reliable satellite measurements began. In mid-September, Arctic sea ice had declined to 3.41 million square kilometers (1.32 million square miles), much less than the previous record of 4.17 million square kilometers (1.61 million square miles) in 2007. Melting sea ice feeds itself. With less ice to reflect sunlight, the dark sea surface absorbs more energy. Less ice survives year to year; new ice is thinner and melts faster.

A few years previously, scientific forecasts had anticipated an ice-free Arctic Ocean in summer by the middle of the twenty-first century, but after several years of rapid ice melt, those projected dates were being moved up to as early as 2020. "It's an example of how uncertainty is not our friend when it comes to climate-change risk," said Michael E. Mann, a climate scientist at Pennsylvania State University. "In this case, the models were almost certainly too conservative in the changes they were projecting, probably because of important missing physics" (Gillis 2012).

## *LAND OF MELTING ICE AND BURNING TUNDRA*

As tundra and permafrost thaw, stored carbon dioxide and methane are released, further raising the atmosphere's level of greenhouse gases. During

Alaska's unusually warm, dry winter of 2000–2001, the snowless tundra caught fire along Norton Sound, a possible precursor of larger, smoldering fires that could further accelerate global warming. There exists a palatable fear among climate scientists that such "biotic feedbacks" in the oceans and on land could release large amounts of carbon dioxide and methane that are now stored in the Earth.

Additional evidence that the Earth's frozen reaches already are releasing additional greenhouse gases into the air has been provided by M. L. Goulden and colleagues, who studied boreal forests, finding evidence that "provides clear evidence that carbon dioxide locked into permafrost several hundred to 7,000 years ago is now being given off to the atmosphere as warming climate melts the permafrost" (Davis 2001, 270). The same is true in many cases for methane. According to Neil Davis, author of *Permafrost: A Guide to Frozen Ground in Transition*, this "relict" carbon dioxide

> represents a massive source since it is estimated that the carbon dioxide contained in the seasonally and perennially frozen soils of boreal forests is 200 to 500 billion metric tons, enough if all released to increase the atmosphere's concentration of carbon dioxide by 50 percent. Hence, it is possible that the release of carbon dioxide from melting permafrost during warming, or locking it into newly frozen soil during cooling, may accelerate climate change. (Davis 2001, 270)

## *WARMING'S PERVASIVE EFFECTS*

A warming climate is reshaping daily life in the Arctic, including decisions about such basics as which streams or rivers are safe sources of drinking water. The beaver population has risen as the weather warms. "The rodent's defecation is suspected of heightening the risk of giardia, an intestinal infection often referred to as 'beaver fever,' " reported *Mother Jones* (Lee 2011). "In general, people could drink from [the creeks and rivers] freely. Now they have beavers defecating into the river," said Michael Brubaker, director of the health consortium's Center for Climate and Health ("Climate Change Puts" 2011).

Rising temperatures enhance the probability of disease in many fish on which Native peoples depend. For example, "Chinook salmon are now commonly infected with a single-celled parasite previously unknown in Alaskan salmon" (Lynn et al. 2013, 550). Salmon, a cold-water fish, are very sensitive to warming water, which debilitates and often makes them ill. In the summer of 2003, "salmon paused their upstream migration, remaining below the Bonneville Dam on the Columbia River, for four weeks until temperatures cooled," noted one account (Cozzetto 2013, 576). Warmer water is causing mussels to disappear in parts of Pacific Northwest waters. Rising ocean acidity provoked by rising levels of carbon dioxide in air and water also is imperiling shellfish. Along some parts of the Pacific Coast, shellfish have stopped reproducing because seawater contains too much carbon dioxide and has become unnaturally acidic.

Far to the south of Alaska, Crow elders' observations describing rising temperatures and diminishing snowpack in southeastern Montana have been compared to weather observations and fleshed out with their descriptions of effects on people, plants, and animals. For example, with both drought and extreme rainfall events increasing, floods and wildfires (at different times) have become more damaging. With stream flow generally declining (except in occasional floods), the distribution of fish on which Native people depend has become more precarious. "Ground [that] used to be covered in deep snow and stayed frozen from November to March [often is now] bare grass for much of the winter" (Doyle et al. 2013, 645).

In Crow country, berries often bud out earlier and then become vulnerable to sudden freezes in spring, killing the blossoms. Longer autumns mean that buffalo berries dry out before the first frost, making them worthless. Because of warmer winters, trees may emerge from dormancy and then die in sudden subsequent freezes. Increasingly hot late springs and summers have made fasting for a sun dance in late May or June risky for people's health. The average temperature in Hardin, Montana, has risen from 45.6 degrees Fahrenheit in the 1950s to 50.1 degrees Fahrenheit after 2000 (Doyle et al. 2013, 648). Snow used to melt slowly, ensuring summer water supplies. Now, shorter winters and hotter summers threaten water supplies by the hot season.

Drought's effects are pervasive in the U.S. Southwest. For example, "Our 30,000-acre reservation is pretty dry because of drought," said Lawrence Snow, land resources manager for Utah's Shivwits Band of Paiutes. "Wildfires in the last decade have burned half our acreage and changed the landscape. We've got less trees, and bark beetles are trying to kill off the ones we do have. Once the fires happened and took out the ground cover, major storms brought big flooding. Seasons have changed; winter hangs around now until May. It's weird the way the weather is changing" (Allen 2012).

Persistent drought in the southern tier of states is forcing Navajos to travel longer distances for potable water and forcing the early sale of livestock. Premature snowmelt on the Pine Ridge provoked by temperature spikes as high at 70 degrees Fahrenheit during what used to be prethaw seasons have caused early flooding, as "emergency drinking water and supplies had to be delivered to stranded residents" (Cozzetto et al. 2013, 578). Meanwhile, tribes in coastal Louisiana are losing land to rising sea levels, subsiding land, and erosion.

"Perhaps among the worst of those impacts," wrote Terri Hansen (2014) for the Indian Country Today Media Network,

> are the runaway sand dunes it has unleashed, which extend over one-third the 27,000-square-mile [Navajo] reservation. During the 1996–2009 drought period the extent of dune fields increased by some 70 percent. These dunes are moving at rates of approximately 35 meters per year, covering houses, burying cars and snarling traffic, degrading grazing and agricultural lands, contributing to the loss of rare and endangered native plants, and when they occur

contributing to poor air quality, a serious health concern for many of the reservation's 173,667 residents.

## *SWEATING IN IQALUIT*

July 29, 2001, was a very warm day in Iqaluit, on southern Baffin Island, capital of semisovereign Inuit nation of Nunavut in the Canadian Arctic. The bizarre weather was the talk of the town. The urgency of global warming was on everyone's lips. Even as U.S. President George W. Bush faulted global warming for lacking "sound science," the temperature hit 25 degrees Celsius in a community that nudges the Arctic Circle, following a string of days that were nearly as warm. It was the warmest summer anyone in the area could remember. Travelers joked about forgetting their shorts, sunscreen, and mosquito repellant—all now necessary equipment for a globally warmed Arctic summer.

In Iqaluit (pronounced "Eehalooeet"), a warm, desiccating northwesterly wind raised whitecaps on nearby Frobisher Bay and rustled carpets of purple saxifrage flowers on bay-side bluffs as people emerged from their overheated houses (which have been built to absorb every scrap of passive solar energy), swabbing ice cubes wrapped in hand towels across their foreheads. The high temperature, at 78 degrees Fahrenheit, was 30 degrees above the July average of 48, comparable to a 110- to 115-degree day in New York City or Chicago. The wind raised eddies of dust on Iqaluit's gravel roads as residents swatted slow, corpulent mosquitoes.

In the Eskimo village of Kaktovik, Alaska, on the Arctic Ocean roughly 250 miles north of the Arctic Circle, a robin built a nest in town during 2003—not an unusual event in more temperate latitudes but quite a departure from the usual in a place where, in the Inupiat Eskimo language, no name exists for robins. Some residents of Baker Lake, Nunavut, 1,330 kilometers west of Iqaluit, spotted magpies flitting around town in May 2006. These scavengers, a relative of the crow, had never been seen in Nunavut before. The magpies are not expected to become permanent residents, however, even if the climate warms, because they roost in trees. The tundra has no trees. In the Okpilak River valley, which has heretofore been too cold and dry for willows, they are sprouting profusely. Never mind the fact that in the Inupiat language, *Okpilak* means "river with no willows." Three kinds of salmon have been caught in nearby waters in places where they were once unknown.

Correspondent Jerry Bowen, appearing on the *CBS Morning News* on August 29, 2002, from Barrow, Alaska, the northernmost town in Alaska, quoted Simeon Patkotak, a native elder, as saying residents there had just witnessed their first mosquitoes. Ice cellars carved out of permafrost were melting as well, forcing local Native people to borrow space in electric freezers for the first time to store whale meat. Average temperatures in Barrow have risen 4 degrees Fahrenheit during the past 30 years (Bowen 2002). The average date on which the last snow melts in Barrow in the spring or summer has receded about 40 days between 1940 and the years after 2000, from early July to, in some years, as early as mid- or late May (Wohlforth 2004, 27).

"We have never seen anything like this. It's scary, *very* scary" said Ben Kovic, Nunavut's chief wildlife manager. "It's not every summer that we run around in our T-shirts for weeks at a time" (Johansen 2001, 18). At 11:30 A.M. on a Saturday in late July, Kovic was sitting in his backyard, repairing his fishing boat, wearing a T-shirt and blue jeans in the warm wind, with many hours of Baffin's 18-hour July daylight remaining. On a nearby beach, Inuit children were building sandcastles with plastic shovels and buckets, occasionally dipping their toes in the still-frigid seawater.

Addressing a U.S. Senate Commerce Committee hearing on global warming on August 15, 2004, Sheila Watt-Cloutier, then president of the Inuit Circumpolar Conference, said, "The Earth is literally melting. If we can reverse the emissions of greenhouse gases in time to save the Arctic, then we can spare untold suffering." She continued, "Protect the Arctic and you will save the planet. Use us as your early-warning system. Use the Inuit story as a vehicle to reconnect us all so that we can understand the people and the planet are one" (Pegg 2004).

The Inuits' ancient connection to their hunting culture may disappear within her grandson's lifetime, Watt-Cloutier said. "My Arctic homeland is now the health barometer for the planet." Committee Chair John McCain (R–AZ), who would later become the Republican candidate for U.S. president, said a recent trip to the Arctic showed him that "these impacts are real and consistent with earlier scientific projects that the Arctic region would experience the impacts of climate change at a faster rate than the rest of the world. We are the first generation to influence the climate and the last generation to escape the consequences," McCain said (Pegg 2004).

Warming has extended to all seasons. In Iqaluit, for example, thunder and lightning used to be extremely rare. The world of the Arctic is changing rapidly before indigenous peoples' eyes. Lightning strikes, for example, ignite fires on the tundra just as in forests; more than 20,000 were recorded on Alaska's North Slope in 2007, a period of record-high temperatures and widespread drought in the area. One strike torched a fire near the Anaktuvuk River. By the time its embers were snuffed out by falling snow in early October, it had scorched an area the size of New York City. According to the U.S. Bureau of Land Management, the frequency of lightning on the North Slope has increased by a factor of 10 in 10 years. Warming brings more lightning, and lightning-sparked fires add carbon dioxide to the air; tundra stores 14 percent of the world's surface soil carbon (Qiu 2009, 34). Fires also accelerate the thawing of permafrost, adding even more carbon dioxide and methane to the atmosphere.

The same day that the high hit 78 degrees Fahrenheit in Iqaluit, the forecast for Yellowknife in the Northwest Territories called for a high of 85 degrees with scattered thunderstorms—a typical summertime forecast in Chicago or New York City. During the summer of 2004, highs in the 80s became routine in Fairbanks, Alaska. The winters of 2000–2001 and 2001–2002 in Iqaluit were notable for liquid precipitation (freezing rain) in December. A few years later, thunderstorms bearing heavy rain were reported during brief temperature spikes even in

February across Alaska's North Slope, over and near Iqaluit, and along the southwestern coast of Greenland.

The harbor ice at Iqaluit did not form in 2000 and 2001 until late November or December, five or six weeks later than usual. In later years, the late onset of ice ceased to be unusual. The ice also breaks up earlier in the spring—sometimes in May, in places that once were icebound, into early July. Early in January 2004, Watt-Cloutier wrote from Iqaluit on Baffin Island that Frobisher Bay had just frozen over for the season at a record-late date:

> We are finally into very "brrrrrr" seasonal weather and the Bay is finally freezing straight across. At Christmas time the Bay was still open and as a result of the floe edge being so close we had a family of Polar bears come to visit the town a couple of times. Also in Pangnirtung [north of Iqaluit] families from the outpost camps came into town for Christmas by boat! Imagine that the ice was not frozen at all there by Christmas. But this week our temperatures with the wind-chill reached minus 52, so we are pleased. (Watt-Cloutier, personal communication, 2004)

Watt-Cloutier looked out at the waters of Frobisher Bay two weeks after she had represented the Inuit Circumpolar Conference at the Conference of Parties to the United Nations Framework Convention on Climate Change in Milan, Italy, where she said, in part,

> Talk to hunters across the North and they will tell you the same story—the weather is increasingly unpredictable. The look and feel of the land is different. The sea-ice is changing. Hunters are having difficulty navigating and traveling safely. We have even lost experienced hunters through the ice in areas that, traditionally, were safe! Our Premier, Paul Okalik, lost his nephew when he was swept away by a torrent that used to be a small stream. The melting of our glaciers in summer is now such that it is dangerous for us to get to many of our traditional hunting and harvesting places. . . . Inuit hunters and elders have for years reported changes to the environment that are now supported by American, British and European computer models that conclude climate change is amplified in high latitudes. (Watt-Cloutier 2003)

Snows in Iqaluit also now tend to be heavier and wetter than previously. Winter cold spells, which still occur, have generally become shorter, according to long-time residents of the area.

Climate change has been rapid and is easily detectable within a single human lifetime. "When I was a child," said Watt-Cloutier, "we never swam in the river [Kuujjuaq] where I was born [in Nunavik, northern Quebec]. Now kids swim there all the time in the summer." Cloutier does not remember even having worn short pants as a child (Johansen 2001, 19).

Some of the rivers to which Arctic char return for spawning have dried up, according to Kovic; the summer of 2001 brought drought as well as record warmth to Iqaluit and its hinterland. Flying above glaciers in the area, Kovic has noticed that their coloration was changing. "The glaciers are turning brown," he said, speculating that melting ice may be exposing debris and that air pollution from southern latitudes may be a factor. Some ringed seals with little or no hair have been caught, said Kovic. Asked why seals have been losing their hair, Kovic answered, "That is a big question that someone has to answer" (Johansen 2001, 19).

As a wildlife officer, Kovic sees changes that alarm him. Polar bears, for example, are becoming shore dwellers rather than ice dwellers, sometimes with dire consequences for unwary tourists. In Resolute Bay, bears attracted by the smell of seal meat have been known to chase children on their way to school. Some local Inuit hunters have turned profits in the thousands of dollars by leading tourists on hunts for land-bound, hungry bears that local people have come to regard as a dangerous nuisance.

Rapid warming has extended into midwinter. During the first week of January 2011, record warmth in Iqaluit drove temperatures above the freezing point, an extreme rarity there in midwinter that created a debilitating, muddy mess. After that, everything flash froze into dangerous sheets of ice. The temperature at one point was 68 degrees Fahrenheit above average. In New York City, a similar departure from average would have produced a temperature of 100 degrees Fahrenheit on New Year's Eve. A high of 35 degrees Fahrenheit in the first week of January was an all-time record. Nearby, the village of Pangnirtung hit 46 degrees Fahrenheit on January 4. "The normal around this time of year is around −22 C (−8 F)," Yvonne Bilan-Wallace, a meteorologist with Environment Canada, told the *Nunatsiaq News* in Iqaluit. "So yeah, you're way above normal" (Windeyer 2011). "Accustomed to blizzards, Iqaluit road crews are finding it challenging to keep the streets passable," reported the Indian Country Today Media Network. "When rain covered the previous day's sand, then froze, they couldn't put down more sand until they had chipped the ice away with a grader" ("Nunavut Capital" 2011).

In the Inuit village of Coral Harbour in the Canadian Arctic near the mouth of Hudson Bay, temperatures also exceeded the freezing point in January 2011 for the first time on record during that month. Hudson Bay could be ice free by the middle of this century. "We're starting to warm up very, very fast," said Peter Scott, scientific coordinator of the Churchill Northern Studies Center ("Hudson Bay" 2001). Scott said that climate change is bringing some species to the Arctic that have never been seen there before. "We may get southern animals. We're seeing a lot of moose now. . . . We now have moose in Churchill, where 20 years ago moose virtually didn't exist" ("Time to Act" 2001).

In the Arctic, wrote Martin P. Tingley and Peter Huybers (2013, 201) in *Nature*,

> The summers of 2005, 2007, 2010 and 2011 were warmer than those of all prior years back to 1400 in terms of the spatial average. The summer of 2010

> was the warmest in the previous 600 years in western Russia and probably the warmest in western Greenland and the Canadian Arctic as well. These and other recent extremes greatly exceed those expected from a stationary climate.

Nick Jans (2008, 150, 153), writing in *The Last Polar Bear*, sketched the ground-level effects of climate change in Barrow:

> A sharp, insistent wind sifts a rolling haze of snow along the ground, but for mid-December the temperature is positively balmy: plus 19 degrees Fahrenheit. A generation ago, 20 or even 30 below zero would have been the norm. The last few days amount to more than some freakish warm spell; this year, the sea ice has yet to become shorefast. Not so many years ago, solid freeze-up occurred by the end of October, without fail. Now it's six, seven, even ten weeks later, and the ice proves thinner and less stable. The elders shake their heads, bemused as if palm trees had sprouted on the shores of the Beaufort Sea. They're beyond debating climate change. It's staring them right in the face.

In the Canadian Inuit town of Inuvik, 90 miles south of the Arctic Ocean, the temperature rose to 91 degrees Fahrenheit on June 18, 1999, a type of weather unknown to living memory in the area. "We were down to our T-shirts and hoping for a breeze," said Richard Binder, 50, a local whaler and hunter. Along the MacKenzie River, according to Binder, "hillsides have moved even though you've got trees on them. The thaw is going deeper because of the higher temperatures and longer periods of exposure." In some places near Binder's village, the thawing earth has exposed ancestral graves, and remains have been reburied (Sudetic 1999, 106). Some hunters say that seals have moved farther north, killer whales are eating sea otters, and beaver are proliferating, something that would not happen if rivers and ponds were freezing to usual depths.

## *THE HUNTER'S LIFE IN TODAY'S ARCTIC*

Coastal hunters above the Arctic Circle in Alaska say that they are definitely seeing a trend: the ice regularly comes a month later than it did 20 years ago and roughly two months later than it did 30 years ago. Ice also breaks up earlier than previously, so hunting seasons are becoming shorter. The Arctic's rapid thaw has made hunting, never a safe or easy way of life, even more difficult and dangerous. Hunters in and around Iqaluit say that the weather has been seriously out of whack since roughly the mid-1990s. Simon Nattaq, an Inuit hunter, fell through unusually thin ice and became mired in icy water long enough to lose both his legs to hypothermia, one of several injuries and deaths reported around the Arctic recently due to thinning ice.

Pitseolak Alainga, another Iqaluit-based hunter, said that climate change compels caution. One must never hunt alone, he said (Nattaq had been hunting by himself). Before venturing onto ice in fall or spring, hunters should test its stability with a harpoon, he said. Such tests are especially important when a hunter travels in a snowmobile. In older times, sled dogs could sense thin ice, something a machine can't do.

Alainga knows the value of safety on the water. His father and five other men died in late October 1994 after an unexpected ice storm swamped their hunting boat. The younger Alainga and one other companion barely escaped death in the same storm. He believes that more hunters are suffering injuries not only because of climate change but also because basic survival skills are not being passed from generation to generation as in years past, when most Inuit lived off the land.

Steven Kooneeliusie and the other Inuit who hunt caribou, seal, and other animals say the signs of a gradual increase in temperatures are everywhere around them. "When I went hunting years ago, I used to wear a full-length caribou skin coat, but now I just wear a light parka. It is so hot these days my snowmobile often overheats," Kooneeliusie said in the small town of Pangnirtung, about 100 miles north of Iqaluit, astride the Arctic Circle (Ljunggren 2000).

Sustainable Development Minister Peter Kilabuk, based in the Nunavut capital of Iqaluit, grew up in Pangnirtung. "I know when I was probably eight or ten [years of age] the ice wouldn't go out until July, sometimes not until the second week of July. But over the last few years we've seen the ice go out as early as May," he said. (In some years, such as 2001, the ice still has departed in late June.) "To us, the effects are real. Climate change is here and it's a real cause for concern." Grizzly bears and wolverines have been moving north, said Francois Rainville at Environment Canada's office in Iqaluit. "There are insects and birds which have not been seen here before. There is an impact. People are seeing change," he said. In 1999, one Iqaluit woman reported seeing a robin (Ljunggren 2000).

Pangnirtung is the main gateway to Auyuittuq National Park, the most northerly national park in North America. (Auyuittuq, pronounced "ow-you-ee-tuk," an Inuit word that is now out of date, means "the land that never melts"). Established in 1972, the park covers 19,500 square kilometers of deep mountain valleys, dramatic fjords, ancient glaciers, and spiny peaks. Auyuittuq is famed for its enormous glaciers. Local people say that even these glaciers are slowly melting. "The glaciers have receded over the last 10 years, and the ice is much worse," hunter Solomon Nakoolak said (Ljunggren 2000).

"If you look at weather patterns for the next 40 years, you'll see a significant warming in the Eastern Arctic, but mainly in the winter," said Brian Paruk, a meteorologist with Environment Canada's Arctic Weather Centre (George 2000). He said that the two main influences on Arctic weather to date have been the "incredible retreat" of the sea ice and the resulting increase in precipitation. More open water produces more rain. Due to the lower temperature of this open water, summer temperatures have declined in many coastal areas (particularly along the Davis Strait) as they have risen inland.

### *A WINTER WITHOUT WALRUS*

Hunting, including that of walrus, is still key to the survival of Native peoples across the Arctic. Imported food is not only culturally out of character but also so expensive that few Natives can afford it. Wild swings in weather and ice conditions have devastated harvests of walrus, the main subsistence staple in some Alaskan Native villages. In recent years, ice, the primary habitat of walrus, has been too scarce—except, as in the winter of 2012–2013, when the weather was so cold and the ice so thick that hunters could not get out often enough to provide for their families.

Food supplied by airplane to the one small grocery store in Gambell, Alaska (in extreme northwestern Alaska within sight of Siberia's Chukchi Peninsula), is no alternative. A single chicken cost $25. Families that once put 20 to 25 walrus in their meat lockers have been storing only a handful in recent years because of retreating ice. "If this continues, we will seriously starve," said Jennifer Campbell, in 2013 a 38-year-old mother of five whose family caught only two walrus in the autumn of 2013, down from about 20 previously. The Yupik Eskimos harvest the walrus from passing ice floes in May and June using small boats. In 2013, however, the hunt lasted only a week. "The ice was so bad we couldn't get out," said Brian Aningayou, a hunter who took only four walrus and who feeds an extended family of 40 (Carlton 2013, A-4). He has been shooting birds, which provide little meat compared to a walrus, which can be the size and weight of a small car.

With ice receding hundreds of miles offshore of Alaska and Russia during the late summer of 2007, walrus gathered by the thousands onshore of Alaska and Siberia. Joel Garlach-Miller, a walrus expert with the U.S. Fish and Wildlife Service, said that walrus began to gather onshore late in July, a month earlier than usual. A month later, their numbers had reached record levels from Barrow to Cape Lisburne, about 300 miles southwest on the Chukchi Sea.

Walrus usually feed on clams, snails, and other bottom-dwelling creatures from the ice. In recent years, the ice has receded too far from shore to allow the usual feeding pattern. A walrus can dive 600 feet, but water under ice shelves in late summer is now several thousands of feet deep. The walrus have been forced to swim much farther to find food. Finding less food, they return to their starving calves. More calves also are being orphaned. Russian observers also reported many more walrus than usual onshore (tens of thousands in some areas along the Siberian coast) that would have stayed on the sea ice in earlier times (Joling 2007). Walrus are prone to stampedes once they are gathered in large groups. Thousands of Pacific walrus above the Arctic Circle were killed on the Russian side of the Bering Strait, where more than 40,000 had hauled out on land at Point Schmidt as ice retreated farther north. A polar bear, a human hunter, or noise from an airplane flying close and low can send thousands of panicked walrus rushing to the water.

"It was a pretty sobering year, tough on walruses," said Garlach-Miller. Several thousand walrus died in the late summer of 2007 from internal injuries suffered in stampedes. The youngest and the weakest animals, many of them calves born the previous spring, were crushed. Biologist Anatoly Kochnev of Russia's Pacific Institute of Fisheries and Oceanography estimated that 3,000 to 4,000 walrus out of population of perhaps 200,000 died, or two or three times the usual number hauling out on the shoreline. "We were surprised that this was happening so soon, and we were surprised at the magnitude of the report," said wildlife biologist Tony Fischbach of the USGS (Joling 2007). Walrus lacking summer sea ice that they use to dive for clams and snails may strip coastal areas of food and then starve them in large numbers.

## *AN ORAL HISTORY OF A MELTING WORLD*

Around the Arctic, in Inuit villages now connected by the oral history of traveling hunters as well as e-mail, weather watchers are reporting striking evidence that global warming is an unmistakable reality. Weather reports from the Arctic sometimes read like the projections of the Intergovernmental Panel on Climate Change on fast-forward. These personal stories support the panel's expectations that climate change will be felt most dramatically in the Arctic.

In the 1990s, Alaskan Native weather watchers established websites to share their experiences with climate change: "Turtles appearing for the first time on Kodiak Island, birds starving on St. Lawrence Island, thunder first heard on Little Diomede Island ... snowmobiles falling through the ice in Nenana. ... Already the central Arctic is warming 10 times as fast as the rest of the planet, outpacing even our attempts to describe it" (Frey 2002, 26). During the summer of 2004, several *Vespula intermedia* (yellow-jacket wasps) were sighted in Arctic Bay, a community of 700 people on the northern tip of Baffin Island, at more than 73 degrees north latitude. Noire Ikalukjuaq, the mayor of Arctic Bay, photographed one of the wasps at the end of August. Ikalukjuaq, who said he knew no word in Inuktitut (the Inuit language) for the insect, reported that other people in the community also had seen wasps at about the same time ("Rare Sighting" 2004).

On Earth Day 2000, the Inuit brought their accounts of dramatic warming to urban audiences in southeastern Canada. Inuit leader Rosemarie Kuptana told a press conference in Ottawa that experienced hunters have fallen through unusually thin ice. Three men had recently died this way. Never-before-seen species (including robins, barn swallows, beetles, and sand flies) have appeared on Banks Island in the Arctic Ocean about 800 miles northeast of Fairbanks, Alaska.

Dwane Wilkin, a reporter for the *Nunatsiaq News* in Iqaluit, summarized the consequences of global warming for that area: "The good news is that sailing through the Northwest Passage will finally be a cinch. The bad news? Well, global warming will probably drive musk oxen, polar bears and Peary caribou into extinction. And other species—including humans—will face declining food sources" (Wilkin 1997). Warmer average temperatures north of 60 degrees north latitude may mean that the Arctic Ocean will become ice free much of the year, imperiling populations of walrus and seal that feed on creatures for which ice means food and life.

The bay on which Iqaluit sits is named for Martin Frobisher, an early explorer who sought a "northwest passage" from Europe to Asia. With the polar ice cap in the Canadian Arctic melting, the fabled Northwest Passage is now opening. Commercial shippers are eyeing the prospect of cutting more than 4,000 miles off their usual trip through the Panama Canal, bringing one more finger of the industrial South to the Arctic.

"Our ice is getting thinner and going out. This year, we had very thin ice and it went out real fast," said Inuit elder Alexander Akeya. "It's getting more windy, and we have to hunt farther out than we used to before." Elder Jimmy Toolie had a similar observation: "Today it has changed, the ice moves all the time, and goes away quicker. We used to have it to mid-June. This year it went out in mid-May. Even in the whaling time, it seemed it went away so quick." Another Inuit, Joe Noongwook, agreed: "This is true. We used to have huge icebergs come first before the young ice formed. We used to have thick ice from here to the [Russian] mainland to the north, but now there is thin ice. There is no hardpack ice" (Braem 1997). Several other Inuit said that the ice has been thinning, especially during the past 20 years.

"Those of us who are dependent on Arctic resources know global warming is occurring," said Caleb Pungowiyi, a Yupik Inuit (Eskimo) elder. "It is affecting us, and the impacts in some cases are quite severe" ("Global Warming" 1998). Pungowiyi lives in Savoonga, a village of about 500 people, near Nome. The people of Savoonga make their livings mainly from the sea, harvesting seals, walrus, polar bears, whales, salmon, trout, whitefish, and other marine species.

Yupik elders in the small coastal Alaskan village of Kipnuk believe that the village is sinking due to warming permafrost. Buildings in the village show signs of an unstable ground surface, signs that are consistent with those of an area that sits above thawing land. In Kotzebue, Alaska, the hospital has been relocated because it was sinking into the ground.

Growing numbers of Inuit are suffering allergies from white-pine pollen that recently reached Kuptana's home in Sachs Harbour on Banks Island for the first time. "If this rate of change continues, our lifestyle may forever change, because our communities are sinking with melting permafrost and our food sources are . . . more difficult to hunt" (Duffy 2000b, A-13). Sachs Harbour itself is slowly sinking during the summer into a muddy mass of thawing permafrost. Born in an igloo, Kuptana has been her family's weather watcher for much of her life (she was 54 years of age in 2008). Her job was to scan the morning clouds and test the wind's direction to help the hunters decide whether to go out and what everyone should wear. Now she gathers observations for international weather-monitoring organizations.

"What happens in the Arctic environment is what is in store for other regions of the world," said Kuptana, who grew up in Sachs Harbour. "We can't read the weather like we used to" (Duffy 2000b). She said that autumn freezes now occur a month later than earlier; spring thaws come earlier as well.

Residents of Sachs Harbour still suffer through winters that most people from lower latitudes would find chilling, with temperatures of minus 40 degrees

Fahrenheit or lower. While such temperatures were once common, however, they now are rare. "The permafrost is melting at an alarming rate," said Kuptana. Foundations of homes in Sachs Harbour have been cracking and shifting. Kuptana expressed fear that the community will soon slide into the Beaufort Sea because once-solid mud is now thawing earlier and freezing later. "What's scary is the uncertainty," she said. "We don't know when to travel on the ice and our food sources are getting further and further away." Tracking animals on receding ice and mushy ground poses new problems for Inuit hunters, some of whom have been reporting sunburns for the first time. "We have no other sources of food, the people in my community are completely dependent on hunting, trapping and fishing," said Kuptana. "We have no means of adapting to a different environmental reality, and that is why our situation is so critical" (Herbert 2000).

At Sachs Harbour, sea ice is thinner and now drifts far away during the summer, taking with it the seals and polar bears on which the village's Inuit residents rely for food. Many young seals are starving to death because prematurely melting sea ice separates them from their mothers. In the winter, the sea ice often is thin and broken, making travel dangerous for even the most experienced hunters. In the fall, storms have become more frequent and often more severe, making boating difficult. Thunder and lightning have been seen for the first time, arriving with another type of weather that is new to the area: dousing summer rainstorms.

"When I was a child, I never heard thunder or saw lightning, but in the last few years we've had thunder and lightning," Kuptana said. "The animals really don't know what to do because they've never experienced this kind of phenomenon" ("Thunderstorms" 2000). Kuptana added, "Our way of life is being permanently altered" (Knight 2000). Andy Carpenter of Sachs Harbour observed that "different kinds of birds are coming up. There's other species of fish that we've never gotten before, like the pink salmon" ("Scientists Look" 2001).

## *THE INUIT WORLD TURNED UPSIDE DOWN*

By the winter of 2002–2003, a warming trend was forcing hockey players in Canada's far north to seek rinks with artificial ice. Canada's *Financial Post* reported, "Officials in the Arctic say global warming has cut hockey season in half in the past two decades and may hinder the future of development of northern hockey stars such as Jordin Tootoo" ("Ice a Scarce Commodity" 2003). According to the *Financial Post* report, hockey rinks in northern communities were raising funds directed toward the installation of cooling plants to create artificial ice because of the reduced length of time during which natural ice was available. In Rankin Inlet, Tootoo's hometown on Hudson Bay in Nunavut, a community of 2,400 residents installed artificial ice during the summer of 2003 after eight years of lobbying for the funds ("Ice a Scarce Commodity" 2003).

Hockey season on natural ice, which in the 1970s ran from September until May, often now begins around Christmas and ends in March, according to Jim MacDonald, president of Rankin Inlet Minor Hockey. "It's giving us about three months of hockey. Once we finally get going, it's time to stop. At the beginning of our season, we're playing teams that have already been on the ice for two or three months," MacDonald said. According to Tom Thompson, president of Hockey Nunavut, there are about two dozen natural ice rinks in tiny communities throughout the territory but only two with artificial ice, Thompson said. Both are in the capital, Iqaluit ("Ice a Scarce Commodity" 2003).

"In my lifetime I will not be surprised if we see a year where Hudson Bay doesn't freeze over completely," said Jay Anderson of Environment Canada. "It's very dramatic. Yesterday [January 6, 2003], an alert was broadcast over the Rankin Inlet radio station warning that ice on rivers around the town is unsafe. The temperature hovered around minus 12 [degrees] C. It's usually minus 37 there at this time of year" ("Ice a Scarce Commodity" 2003).

Along with artificial ice, some Inuit are coming to know another imported generator of greenhouse gases. During 2006, officials in Nunavut authorized the installation of air conditioners in official buildings for the first time because summertime temperatures in some southern Arctic villages have climbed into the 80s in recent years.

The weather is not all that has been changing in Nunavut. The very fabric of Inuit life has changed within two generations. Climate change has arrived as Inuit society changes fundamentally in many ways. Inuit elders describe the coming of *qallunaat*, the non-Inuit money economy, in natural metaphors, as a time when the old world shifted, "they say, like a storm they could not read in the clouds. . . . Against the wind, the people first had to brace themselves, then they had to adapt, then they had to try to stop the wind." The change has come with a stunning swiftness, in some cases within a generation or two. According to one report,

> Jamaise Mike was born in 1928 in an igloo, but now he is sitting in a Kentucky Fried Chicken restaurant in Pangnirtung, Nunavut. "Most elders were born and raised on the land," Mike said, through an interpreter, speaking in the Inuktitut language. "Everything people needed to survive surrounded them on the land. Today it is different, living in a community with a store. Everything is different from when you had to do everything yourself to survive. You depended on yourself. Now, you need money." (Brown 2001, A-1)

As in many other parts of the world, the story of indigenous environmentalism is shaped, first and foremost, by the collision of the money economy and traditional, environmentally based economies. Many of today's Inuit are the first or second generation off the land, having moved to towns at the behest of the Canadian federal government. In some cases, the government's agents systematically killed many of the dogs that pulled their sleds, giving people no choice but to come in from the land. Children were put in Christian boarding schools. "White people from the south were more terrifying than polar bears," said one Inuit (Brown 2001, A-1).

The collision of cultures can be tragic. Where a few decades ago, consumption of alcohol was rare, Baffin Island Inuit now have Canada's highest rates of drunkenness. Less than 50 years ago, suicide was virtually unknown among the Inuit of Canada's Arctic. Now they are killing themselves at a rate six times the Canadian national average. In 1999, 58 of Nunavut's 27,000 people committed suicide. Fifty-two were by hanging and six by firearms. Fifty-seven of them were Inuit.

The population of Iqaluit jumped from about 3,500 to 6,500 in less than three years between 1998 and 2001. Substantial suburban-style houses with mortgages worth hundreds of thousands of dollars sprang up around town, rising on stakes sunk into the permafrost and granite hillsides. In other areas, ranks of walk-up apartments marched along the high ridges above Frobisher Bay. Every ounce of building material had been imported from thousands of miles away. Inuit have been moving into the town from the backcountry. The town was awash in a sea of children.

Urbanization has caused many Inuit to lose their cultural bearings, including their hunting skills. Within two or three generations, many Inuit have become urbanized as the tendrils of industrial life infiltrate the Arctic. Where visitors once arrived by dogsled or sailing ship, they now stream into Iqaluit's small but very busy airport on Boeing 727s in which half the passenger cabin has been sequestered for freight. With no land-surface connections to the outside, freight as large as automobiles is sometimes shipped to Iqaluit by air.

People in Iqaluit subscribe to the same cable television services (as well as the Internet) available in the "South." Bart Simpson is a well-known personage in Iqaluit, where some homes have sprouted satellite dishes. Iqaluit also now hosts a large supermarket of a size that matches stores in larger urban areas, except that the prices are three to four times higher than in Ottawa or Omaha. If one can afford the bill, however, orange/mango/grapefruit juice and ready-cooked barbecue buffalo wings, as well as many other items of standard "southern" fare, are readily available.

As the economic system of the "South" sets down roots, the Inuit sense a more acute need to preserve traditions that are slipping away. "We live in wooden houses, drive Jeep Cherokees, and fly in jumbo jets all over the world. But we are still Inuit. It is our spirit, our inner being, that makes us Inuit," said John Amagoalik, former chief commissioner of the Nunavut Implementation Commission (Brown 2001, A-1).

## *THE ARCTIC OCEAN IS ACIDIFYING*

The Arctic Monitoring and Assessment Program reported in 2013 that the top of the world is not only rapidly melting but also acidifying due to rising levels of carbon dioxide. The Arctic Ocean is notably vulnerable, according to Rashid Sumaila of the University of British Columbia, who was among 60 scientists who took part in the study. Aboriginal peoples living above the Arctic Circle are affected. "Aboriginal people actually depend a lot on the living sources in the

Arctic," Sumaila told CBC News. "They are very connected to the system, and they will be the first ones to be hit by this" ("Climate Change: Arctic" 2013). While acidity in oceans worldwide has increased 30 percent in 200 years, levels are rising even more sharply in the Arctic Ocean because it receives more freshwater than do the southerly latitudes. Arctic water also is relatively cold and therefore is less able to neutralize the increased acidity.

"Because Arctic marine food webs are relatively simple, Arctic marine ecosystems are vulnerable to change when key species are affected by external factors," the monitoring program said in a statement. "Arctic marine ecosystems are highly likely to undergo significant change due to ocean acidification." As a result, "marine mammals, important to the culture, diets and livelihoods of Arctic Indigenous Peoples and other Arctic residents could also be indirectly affected through changing food availability," the research team said ("Climate Change: Arctic" 2013).

## *ALASKAN VILLAGES FALL TO ENCROACHING SEAS AS OBAMA ACKNOWLEDGES TOLL*

In November 2013, President Barack Obama's White House held a Tribal Nations Conference that ended with an agreement to address global warming's effects on peoples represented by the National Congress of American Indians (NCAI). During the conference, the Indian Country Media Network reported that "several federal officials noted the severe impacts that climate change has had on American Indians and Alaska Natives" ("Feds Reach Out" 2013). "The health of tribal nations depends on the health of tribal lands. So it falls on all of us to protect the extraordinary beauty of those lands for future generations," Obama said at the Tribal Nations Conference. "And already, many of your lands have felt the impacts of a changing climate, including more extreme flooding and droughts," Obama told the conference ("Feds Reach Out" 2013).

"It is critically important that tribal leaders are at the table because too often, Native voices are left out of federal conversations around mitigating the effects of climate change," the NCAI said. "Indian country faces some of the most difficult challenges stemming from climate change because of the remote location of many tribal lands and, particularly in Alaska, the dependence on the land and animals for subsistence living. NCAI applauds the Administration for this effort and is hopeful that by working together, government-to-government, tribal communities will have the tools necessary to address climate change" ("Feds Reach Out" 2013).

As this conference met, several coastal Alaskan Native villages were declared disaster areas following a series of severe storms. By 2013, more than 30 Alaskan Native villages were facing relocation due to climate change, including Shishmaref and Newtok. "Alaska is seeing all these things the rest of the country hasn't seen yet," said Dr. Jerome Montague, Native affairs and natural resources adviser for the Alaskan Command Joint Task Force (Jessepe 2012). In 2006, to quell erosion, a new seawall was built on the seaward side of the village of Kivalina, Alaska, costing $3 million.

In Kivalina, erosion and flooding have caused sediment to pour into the community water system, requiring filtration to prevent human ingestion of unsanitary water. About 400 people in Kivalina could take only sponge baths and had limited access to laundry for the winter. "As hand-washing and bathing decreased, respiratory and skin diseases increased," health aides said ("Climate Change Puts" 2011). In Point Hope, algae blooms clouded a lake used for drinking water. Kivalina hangs precariously onto a slender, eroding peninsula, as the U.S. Army Corps of Engineers projects that the sea will wipe it off the map by 2025. In addition, the threat of food poisoning increases with rising temperatures. Meat stored in ice cellars carved from melting permafrost can become contaminated with pathogens that cause sickness. "We used to have frozen whale meat and maktuk all winter and summertime, too," said Joe Towksjhea, a Point Hope resident, in the consortium's report. "It is not frozen anymore" ("Climate Change Puts" 2011). According to Sylvester Ayek, who hunts on King Island nestled in the Bering Sea, the greater problem is the dwindling number of animals due to late freeze and early thaw. "What we hunt is usually around ice. And when the ice goes earlier, like these past few years, the game is gone," Ayek told Indian Country Today Media Network. This issue appears to drastically affect villagers, and especially elders, in Point Hope, leading to more reports of malnutrition and anemia. Adding to the lack of available food, the period for rack-drying fish, seal, and caribou has shortened, increasing the likelihood of food-borne illnesses caused by bacteria—not to mention the lingering scent of raw meat in milder temperatures that attracts hungry polar bears ("Climate Change Puts" 2011).

The Indian Country Media Network reported that "the Yup'ik village of Kotlik, Alaska, along with Unalakleet and other predominantly Native communities, were ravaged beginning November 9 [2013] by a series of four storms battered hundreds of miles of Alaska's west coast with near hurricane-force winds, a sea surge as high as nine feet, freezing rain, and snow," Alaska Governor Sean Parnell said. "The storm surge wrecked food supplies" ("Yup'ik" 2013). For a time, the entire town was flooded with seawater and chunks of ice. Rising seas and coastal erosion directly threaten Tuktoyaktuk, a Dene and Inuit community located at the edge of the Arctic Ocean. Ice that once protected the coast has receded out to sea. Extensive erosion washed away the school and has forced the village to relocate many other structures.

These Native villages have been lashed by the sea during increasingly stormy weather for several years. Six hundred people in the Inuit village of Shishmaref on the Chukchi Sea, about 60 miles north of Nome, have been watching their village erode into the sea for several agonizing years. The permafrost that had reinforced its coast is thawing. "We stand on the island's edge and see the remains of houses fallen into the sea," wrote Anton Antonowicz (2000) of the *Daily Mirror*. "They are the homes of poor people. Half-torn rooms with few luxuries. A few photographs, some abandoned cooking pots. Some battered suitcases." Percy Nayokpuk, a village elder, runs the local store, which now perches dangerously close to the edge of the advancing sea. "When I was a teenager, the beach stretched at least 50 yards further out," said Percy. "As each year passes, the sea's approach seems faster" (Antonowicz 2000, 8). Five houses have washed into the sea; the U.S. Army has moved or jacked up others. The villagers have been told they will soon have to move.

Year by year, the hunting season, which depends on the arrival of the ice, starts later and ends earlier. "Instead of dog mushing, we have dog slushing," said Clifford Weyiouanna, 58, a reindeer herder near Shishmaref (Antonowicz 2000, 8). Villagers have been catching fish, such as flounder, which are usually associated with warmer water.

By the summer of 2001, the encroaching sea was threatening a rusty-fuel-tank farm holding 80,000 gallons of gasoline and stove oil. "Several years ago," observed Kim Murphy of the *Los Angeles Times*, "the tanks were more than 300 feet from the edge of a seaside bluff. But years of retreating sea ice have sent storm waters pounding, and today just 35 feet of fine sandy bluff stands between the tanks and disaster" (Murphy 2001, A-1). Seawater is lapping near the town's airport runway, its only long-distance connection to the outside. At last count, three houses had been washed into the sea. Several more are threatened. The town's drinking-water supply also has been inundated by the sea. The sea is eight feet from cutting the town's main road and threatens to wash its dump out to sea.

Shishmaref hunters are being forced to search up to 200 miles from town for walrus because of retreating ice. They also now use boats to hunt seals that they used to track over ice. "This year the ice was thinner, and most of the year at least part of the ice was open. We don't normally see open water in December," said Edwin Weyiouanna, an artist who has lived most of his life on the Chukchi Sea (Murphy 2001, A-1). In earlier years, the sea was more likely to be frozen during much of the stormy winter season. With warming, the erosive, wind-whipped ocean corrodes Shishmaref's waterfront. The town's residents have come to fear the full moon with the unusually high tides it brings. Point Hope, on a spit of land that juts into the Chukchi Sea, has been losing its battle against the sea. Almost $2 million has been spent to shore up its runway (the only connection with the outside) except for an evacuation road restored by the Army Corps of Engineers for $433,000.

By late 2015, U.S. President Barack Obama, with much media fanfare, provided a nudge that may help break the bureaucratic impasse that is condemning Alaska coastal Native villages. In the waning months of his second term, Obama visited Alaska on a "climate change tour," including some of the villages, and pledged aid. Whether the pledge will survive his presidency will be determined after this book sees print.

A *New York Times* account noted that many of the Native people to whom Obama spoke expressed "a hefty dose of skepticism . . . [because in the past they had] often felt ignored, marginalized and victimized by the federal and state governments, and a sense that Mr. Obama's efforts do not come close to meeting their urgent needs" (Davis 2015). Obama announced that the Denali Commission, the federal agency that coordinates government assistance to communities in Alaska, would oversee $2 million in short- and long-term programs to safeguard and repair the coastal villages, and was committing $2 million to such initiatives, including "voluntary relocation efforts, where appropriate." Native people said that $2 million was barely a drop in the proverbial bucket. "What is $2 million going to give us?" said Diane Ramoth of Sewalik, a village of 900 people that has been trying for 10 years to relocate to higher ground. "A dream?" (Davis 2015).

"Sea ice is a pretty sensitive indicator," said Gunter Weller, a professor of geophysics and director of the Center for Global Change and Arctic System Research at the University of Alaska Fairbanks. "It doesn't take much [temperature change] to make a change in the ice" (Kizzia 1998). Weller noted that researchers on the National Science Foundation's ice ship *SHEBA*, north of Alaska, found the polar ice cap to be considerably thinner than previous studies had indicated. They couldn't find an ice floe thick enough to anchor their icebreaker safely. Lack of sea ice causes a feedback warming effect because open ocean absorbs more of the sun's energy than do ice and snow.

Weller said that mean temperatures in the state have increased by 5 degrees Fahrenheit in the summer and 10 degrees Fahrenheit in the winter over the past 30 years. Moreover, the Arctic ice field has shrunk by 40 to 50 percent over the past few decades and has lost 10 percent of its thickness, studies show. "These are pretty large signals, and they've had an effect on the entire physical environment," Weller said (Murphy 2001, A-1).

Increasing coastal erosion isn't limited to Shishmaref; it is now occurring on the shores of the Bering Sea, as increasing storm surges crush ice packs that retreat several weeks earlier they did than 30 years ago. The extent of sea ice in the Arctic has decreased Arctic-wide by an average of 0.35 percent per year since 1979. During the summer of 1998, record reduction of sea-ice coverage was observed in the Beaufort and Chukchi seas (Maslanik et al. 1999). In Tuktoyaktuk, a town near the mouth of the Mackenzie River, several buildings have been lost to the sea by erosion. In addition to a reduction in Bering Sea ice cover, more precipitation is falling in many areas of Alaska. During the fall of 1998, sea ice formed in northern Alaska more than a month later than usual, postponing the annual seal and walrus hunt. Average temperatures at the mouth of the Mackenzie River were 9 degrees Fahrenheit above long-term averages during 1998. Several decades ago, miners in the area deposited toxic wastes in ponds that were expected to remain frozen (and the toxic materials sealed by the permafrost). With warmer temperatures, some of these toxic dumps may thaw and leak.

In Alaska, where roughly 80 percent of glaciers are receding, forests of dead spruce surround Anchorage, a casualty of a spruce-beetle epidemic caused at least in part by rising temperatures, which accelerate the insect's life cycle. Large patches of Alaskan forests have been described as drowning and turning gray as thawing ground sinks under them. Trees and roadside utility poles, destabilized by thawing, lean at crazy angles. The warming has contributed a new phrase to the English language in Alaska: the "drunken forest."

### *HUNGRY POLAR BEARS AND A DYING SEA-ICE ECOLOGY*

The polar bear is one of the most visible casualties of a deteriorating ice ecology in the Arctic. Polar bears find their food supply shrinking with the melting sea ice. In the pre–global warming days, polar bears had their own food sources and usually went about their business without trying to swipe food from humans. Beset by the melting of the food-bearing ice that once supplied them with protein, hungry polar bears are coming into contact with people more frequently. In Churchill, Manitoba, bears waking from their winter's slumber have found Hudson Bay's ice melted. Instead of making their way onto the ice in search of food, the bears walk along the coast until they get to Churchill, where they get in the way of motor vehicle traffic and pillage the town dump. Churchill now has a holding tank for wayward polar bears that is larger than its jail for people, a result of rising temperatures and melting ice (Krajick 2001, 424). Around the Arctic in the fall, polar bears have been observed anxiously pacing the shorelines, waiting for the ice, which is arriving increasingly later.

Canadian Wildlife Service scientists reported during December 1998 that polar bears around Hudson Bay were 90 to 220 pounds lighter than they had been 30 years ago, apparently because earlier ice melting has given them less time to feed on seal pups. When sea ice fails to reach a particular area, the entire ecological cycle is disrupted. When the ice melts, the polar bears can no longer use it to hunt for ring seals, many of which also have died, having had no ice on which to haul out.

During 1999, a hungry polar bear mother and a cub surprised some picnickers just outside the Inuit village of Pangnirtung. The bears were shot after they could not be scared away. On another occasion, a school group from Arctic Bay on the Arctic Ocean at the northern tip of Baffin Island was camped outside of town. The children were playing radios and otherwise making too much noise to notice the polar bear that entered their camp and stuck its head into the tent of their biology teacher. Awakened by the uproar, the group's guide, Inuit Simon Qamanariq, emerged from his tent to find the bear between him and his *qamatiik*, the sledge pulled by his snowmobile, where he kept his rifle. Qamanariq had at hand only a tea kettle, which Simon threw at the polar bear's feet, distracting it. While the bear was playing with the tea kettle, Qamanariq ran for his gun, seized it, and rattled off a few shots, sending the bear running (Rauber n.d.).

### *MOVING BEYOND A POINT OF NO RETURN*

Scientists were shaken by the sudden retreat of the Arctic ice during the summer of 2007, much more than their models had anticipated. "The Arctic is often cited as the canary in the coal mine for climate warming," said Jay Zwally, a climate expert at NASA. "Now as a sign of climate warming the canary has died"

(Kolbert 2007, 44). "We could very well be in that quick slide downward in terms of passing a tipping point," said Mark Serreze, a senior scientist at the National Snow and Ice Data Center, in Boulder, Colorado. "It's tipping now. We're seeing it happen now." Bob Corell, who headed a multinational Arctic assessment, said, "We're moving beyond a point of no return" ("As Arctic Sea Ice Melts" 2008).

By 2007, Baffin Island in the Canadian Arctic had lost half its ice in 50 years as glaciers on its northern mountain ranges eroded. In another 50 years, what remains may be gone, according to research by geological scientist Professor Gifford Miller and colleagues at the Institute of Arctic and Alpine Research at the University of Colorado, Boulder. "Even with no additional warming, our study indicates these ice caps will be gone in 50 years or less," he said (Anderson et al. 2008). In a separate study released in March 2006, Stroeve and her team showed that dwindling Arctic sea ice may have reached "a tipping point that could trigger a cascade of climate change reaching into Earth's temperate regions" ("Arctic Ice Retreating" 2007). "This suggests that current model projections may in fact provide a conservative estimate of future Arctic change, and that the summer Arctic sea ice may disappear considerably earlier than IPCC [International Panel on Climate Change] projections," said Stroeve ("Arctic Ice Retreating" 2007). In the past, low-ice years often were followed by recovery the next year, when cold winters favored accumulation or cool summers kept ice from melting. That kind of balancing cycle stopped after 2002. "If you look at these last few years, the loss of ice we've seen, well, the decline is rather remarkable," said Serreze (Boyd 2002, A-1; Human 2004, B-2).

At the same time, Alaska's boreal forests are expanding northward at a rate of about 100 kilometers per 1-degree-Celsius rise in temperatures. Ice cover on lakes and rivers in the mid- to high northern latitudes now lasts about two weeks less than it did 150 years ago. During late June 2004, as temperatures in Fairbanks, Alaska, peaked in the upper 80s on some days, a flood warning was issued near Juneau, Alaska—not for rainfall but for glacial snowmelt.

Arctic ice provides an important breeding ground for plankton, NASA senior research scientist Josefino Comiso said. Plankton makes up the bottom rung of the ocean's food chain. "If the winter ice melt continues, the effect would be very profound especially for marine mammals," Comiso said. Arctic ice sometimes melts even in winter as water warms, Comiso said. The winter ice season shortens every year ("Arctic Ice Melting" 2006).

## *SALMON DECLINE IN WARMING WATERS*

The decline of Pacific salmon runs suggests that global climate change could devastate fish populations on which millions of Native (and other) people rely for food. In Alaska, fewer salmon have returned from the ocean to spawn at their birthplaces. Those that do return have been smaller than average and have arrived later than usual. The key factor in the decline of salmon runs appears to have been the fact that water temperatures often have been much higher than usual, stressing these cold-water fish.

Salmon are very sensitive to temperature. If the temperature rises 2 degrees Celsius (from a usual late-summer peak of 21 to 23 degrees Celsius), none of the salmon will survive the swim upstream to spawn. At 23 degrees Celsius, most salmon will die of heat prostration. According to a report by the World Wildlife Fund, "While salmon can withstand higher temperatures in summer when food is abundant, in the winter their tolerance drops considerably. As cold-blooded creatures, their metabolism increases in warmer water and keeping up with this high metabolism requires large amounts of food. If sufficient food is not available, salmon can starve" (Mathews-Amos and Berntson 1999).

For Canada's western regions, climate models forecast an increase in precipitation, water runoff, and flooding in winter and a decrease in precipitation and runoff during summer. Higher winter river flows are expected to damage salmon spawning grounds, reduce the survival and growth of fish because of increased stream temperatures, and damage Fraser River salmon due to increased predation by warm-water species (Rolfe 1996). In 1995, the Canadian Department of Fisheries and Oceans blamed a collapse of Fraser River salmon runs on predation by mackerel, which invaded the salmon spawning grounds, along with warmer-than-average ocean waters provoked by El Niño conditions. At the same time (and for the same reasons, according to the Canadian government), the Queen Charlotte chinook salmon runs declined about 80 percent. Warmer seawater decreases the population of krill, which the salmon eat. Anglers also say that sea lice from salmon reared on fish farms are infesting wild salmon, killing them. The fish farms also produce a polluting slurry. In eight of 32 rivers on the west coast of Scotland, salmon were virtually extinct by 2000. Seals are killing salmon in some rivers as well (Buxton 2000, 11).

Rising temperatures in British Columbia's largest sockeye salmon spawning river, the Fraser, provoked a government shutdown of commercial salmon fishing at the height of the season in late September 1998. According to an Environmental News Service dispatch from Victoria, British Columbia (Thomas 1998), above-average temperatures in the river impaired the salmon's swimming and jumping abilities and afflicted them with "proliferating pathogens and an invasion of warm-water predators such as squaw fish." Seawater temperatures 1.5 degrees Celsius above average in the adjacent Georgia Strait could trigger toxic algae blooms (Thomas 1998).

Between one-quarter and two-thirds of salmon returning to various locations in the Fraser River system were dying before they could spawn, many of them from conditions related to rapid warming of their aquatic environment. The water warmed because of below-average snowpack and above-normal temperatures in the interior of British Columbia, the Fraser's drainage basin. A large aluminum smelter also was delivering large amounts of heated water to an upstream tributary of the Fraser, adding yet another human provocation to the salmon's warming environment. Clear-cutting of upstream forests is also blamed for some of the warming of the river's waters. By late July 2006, the snow pack in the Fraser River watershed of British Columbia, which in most years would be feeding into the river and cooling it, had almost entirely melted. Salmon were threatened by a combination of warm water and low river levels.

## *ANNUAL* MANOOMIN *(WILD RICE) HARVEST CANCELED*

The summer crop of *manoomin* ("food that grows on the water") on northern Wisconsin's Bad River Reservation historically has been so important that the annual powwow is named for it. However, in 2012, the annual *manoomin* harvest was canceled for only the second time in history. Myron Burns, a Bad River tribal member, told the Indian Country Today Media Network that there was simply no rice to harvest. "There was nothing but empty hulls," he said (Pember 2012). The harvest also was also cancelled in 2007 because of drought related to extremely low water levels.

The drought–deluge cycle that is a distinct feature of climate change is wreaking havoc on wild rice. While drought wiped out large parts of the wild rice crop in 2007, massive rainstorms drowned the harvest on the Fond du Lac Reservation in 2012. In 2013, however, precipitation closer to average brought the crop back, at least partially. Some harvests around the state were still restricted (LeMay 2013).

Mary Annette Pember (2012), writing for the Indian Country Today Media Network, said that "Manoomin is central to Ojibwe stories and is believed to be the primary reason the tribe migrated west from the St. Lawrence Seaway area centuries ago. They traveled to the place where the food grows on the water, and settling around the Great Lakes. Manoomin is also a long-standing conduit of culture, language and community. The backbreaking harvest brings families together in an ancient labor that offers food for sustenance and spiritual activities. The harvest also provides some with a much-needed boost in income."

In 2012, however, most Native peoples in the region lost their wild rice. Near Duluth, Minnesota, the Fond du Lac Band of Lake Superior Chippewa canceled its wild rice harvest. According to an article in the *Duluth News-Tribune*, better crops are reported farther west and north, where less rain fell during the crucial spring "floating-leaf" stage of the plant. "It almost brought tears to my eyes to see the sloughs like they are" (Pember 2012)—one more signal, among many, that a changing climate is throwing the atmosphere out of kilter in ways that reshape Native American lives.

## *FURTHER READING*

Allen, Lee. 2012. "Southwest Tribes Struggle with Climate Change Fallout." Indian Country Today Media Network, June 14. http://indiancountrytodaymedianetwork.com/article/southwest-tribes-struggle-with-climate-change-fallout-118386

"American Indians Feel the Effects of Climate Change at Higher Rate Than Other Groups." 2011. Indian Country Today Media Network, August 9. http://indiancountrytodaymedianetwork.com/article/american-indians-feel-the-effects-of-climate-change-at-higher-rate-than-other-groups-46365

Anderson, Rebecca K., Gifford H. Miller, Jason P. Briner, Nathaniel A. Lifton, and Stephen B. DeVogel. 2008. "A Millennial Perspective on Arctic Warming from 14-C in Quartz and Plants Emerging from Beneath Ice Caps." *Geophysical Research Letters* 35: L01502. doi:10.1029/2007/GL032057.

Antonowicz, Anton. 2000. "Baking Alaska: As World Leaders Bicker, Global Warming Is Killing a Way of Life." *Daily Mirror*, November 28, 8–9.

"Arctic Ice Melting Rapidly, Study Says." 2006. *New York Times*, September 14. http://www.nytimes.com/aponline/us/AP-Warming-Sea-Ice.html

"Arctic Ice Retreating 30 Years Ahead of Projections." 2007. Environment News Service, April 30. http://www.ens-newswire.com/ens/apr2007/2007-04-30-04.asp

"As Arctic Sea Ice Melts, Experts Expect New Low." 2008. *New York Times*, August 28.http://www.nytimes.com/2008/08/28/science/earth/28seaice.html

Bowen, Jerry. 2002. "Dramatic Climate Change in Alaska." *CBS Morning News*, August 29.

Boyd, Robert S. 2002. "Earth Warming Could Open Up a Northwest Passage." *Pittsburgh Post-Gazette*, November 11, A-1.

Braem, Nicole M. 1997. "Greenpeace Activists Visit Yup'ik and Inupiat Villages in Alaska to Gather Information about Global Warming." *Nunatsiaq News*, August 8. http://www.nunanet.com/~nunat/week/70808.html

Brown, DeNeen L. 2001. "Culture Corrosion in Canada's North; Forced into the Modern World, Indigenous Inuit Struggle to Cope." *Washington Post*, July 16, A-1.

Buxton, James. 2000. "Suspects in the Mystery of Scotland's Vanishing Salmon: Fish Farms, Seals and Global Warming Are All Blamed for What Some See as a Crisis." *Financial Times*, June 13, 11.

Carlton, Jim. 2013. "A Winter without Walrus: Harvesting of Food Staple for Remote Eskimo Villages Plummets; Disaster Declared." *Wall Street Journal*, October 4, A-4.

"Climate Change: Arctic Ocean Is Acidifying Fast, Study Shows." 2013. Indian Country Today Media Network, May 15. http://indiancountrytodaymedianetwork.com/2013/05/15/climate-change-arctic-ocean-acidifying-fast-study-shows-149369

"Climate Change Puts Health of Arctic Villagers on Thin Ice." 2011. Indian Country Today Media Network, March 7. http://indiancountrytodaymedianetwork.com/article/climate-change-puts-health-of-arctic-villagers-on-thin-ice-21391

Cozzetto, K., K. Chief, K. Dittmer, M. Brubaker, R. Gough, K. Souza, F. Ettawageshik, S. Wotkyns, S. Opitz-Stapleton, S. Duren, and P. Chavan. 2013. "Climate Change Impacts on the Water Resources of American Indians and Alaska Natives in the U.S." *Climatic Change* 120: 569–84.

Davis, Julie Hirschfeld. 2015. "Obama Takes Climate Message to Alaska, Where Change Is Rapid." *New York Times*, September 5. http://www.nytimes.com/2015/09/03/us/politics/obama-takes-climate-message-to-alaska-where-change-is-rapid-in-alaska.html

Davis, Neil. 2001. *Permafrost: A Guide to Frozen Ground in Transition*. Fairbanks: University of Alaska Press, 2001.

Doyle, John T., Margaret Hiza Redsteer, and Margaret J. Eggers. 2013. "Exploring Effects of Climate Change on Northern Plains American Indian Health." *Climatic Change* 120: 643–55.

Duffy, Andrew. 2000a. "Global Warming Causing Arctic Town to Sink, Says Inuit Leader 'Warning Signal.'" *Montreal Gazette*, April 18. http://www.climateark.org/articles/2000/2nd/glwatosi.htm

Duffy, Andrew. 2000b. "Global Warming Why Arctic Town Is Sinking: Permafrost Is Melting under Sachs: Inuit Leader." *Montreal Gazette*, April 18, A-13.

Eilperin, Juliet. 2012. "Alaskan Arctic Villages Hit Hard by Climate Change." *Washington Post,* August 5. http://www.washingtonpost.com/national/health-science/alaskan-arctic-villages-hit-hard-by-climate-change/2012/08/05/e9dbd4a6-d5b0-11e1-a9e3-c5249ea531ca_print.html

"Feds Reach Out to Natives on Climate Change at Tribal Nations Conference." 2013. Indian Country Media Network, November 17. http://indiancountrytodaymedianetwork.com/2013/11/17/feds-reach-out-natives-climate-change-tribal-nations-conference-152290

"First Stewards Resolution Asks U.S. Congress to Formally Recognize Ecological Knowledge of Coastal Indigenous People." 2012. Indian Country Today Media Network, September 30. http://indiancountrytodaymedianetwork.com/article/first-stewards-resolution-asks-u.s.-congress-to-formally-recognize-ecological-knowledge-of-coastal-indigenous-people-136802

Frey, Darcy. 2002. "George Divoky's Planet." *New York Times Sunday Magazine*, January 6, 26–30.

George, Jane. 2000. "Sierra Club Focuses on Arctic Global Warming." *Nunatsiaq News*, January 31. http://www.nunatsiaq.com/archives/nunavut000131/nvt20121_04.html

Gillis, Justin. 2012. "Sea Ice in Arctic Measured at Record Low." *New York Times,* August 27. http://www.nytimes.com/2012/08/28/science/earth/sea-ice-in-arctic-measured-at-record-low.html

"Global Warming Worries Native Americans." 1998. *Deseret News*, November 27. http://www.desnews.com/cit/071en14f.htm

Hansen, Terri. 2014. "Climate Disruptions Hitting More and More Tribal Nations." Indian Country Today Media Network, May 7. http://indiancountrytodaymedianetwork.com/2014/05/07/climate-disruptions-hitting-more-and-more-tribal-nations-154747

Herbert, H. Josef. 2000. "Inuit Say They Are Witness to Global Warming in the Arctic." *Milwaukee Journal-Sentinel*, November 19. http://www.jsonline.com/alive/news/nov00/warming20111900.asp

"Hudson Bay Ice-Free by 2050, Scientists Say." 2001. Canadian Broadcasting Corporation, March 14. http://north.cbc.ca/cgi-bin/templates/view.cgi?/news/2001/03/14/14hudsonice

Human, Katy. 2004. "Disappearing Arctic Ice Chills Scientists; A University of Colorado Expert on Ice Worries that the Massive Melting Will Trigger Dramatic Changes in the World's Weather." *Denver Post*, October 5, B-2.

"Ice a Scarce Commodity on Arctic Rinks: Global Warming Blamed for Shortened Hockey Season." 2003. *Financial Post*, January 7, A-3.

"Indigenous Alaskans Shed Light on Climate Change." 2011. Indian Country Today Media Network, September 14. http://indiancountrytodaymedianetwork.com/article/indigenous-alaskans-shed-light-on-climate-change-53921

Jans, Nick. 2008. "Living with Oil: The Real Price." In *The Last Polar Bear: Facing the Truth of a Warming World*, edited by Steven Kazlowski, 147–61. Seattle: Braided River.

Jessepe, Lorraine. 2012. "Alaskan Native Communities Facing Climate-Induced Relocation." Indian Country Today Media Network, June 21. http://indiancountrytodaymedianetwork.com/article/alaskan-native-communities-facing-climate-induced-relocation-119615

Johansen, Bruce E. 2001. "Arctic Heat Wave." *The Progressive*, October, 18–20.

Joling, Dan. 2007a. "Thousands of Pacific Walruses Die; Global Warming Blamed." Associated Press, December 14.

Joling, Dan. 2007b. "Walruses Abandon Ice for Alaska Shore." *Washington Post*, October 4. http://www.washingtonpost.com/wp-dyn/content/article/2007/10/04/AR2007100402299_pf.html

Kerr, Richard A. 2012. "Ice-Free Arctic Sea May Be Years, Not Decades, Away." *Science* 337 (September 28): 1591.

Kizzia, Tom. 1998. "Seal Hunters Await Late Ice." *Anchorage Daily News*, November 28. http://www.adn.com/stories/T98112872.html

Knight, Danielle. 2000. "Inuit Tell Negotiators of Climate Change Impact." Interpress Service, November 16.

Kolbert, Elizabeth. 2007. "Testing the Climate." *The New Yorker*, December 24 and 31, 43–44.

Krajick, Kevin. 2001. "Arctic Life, on Thin Ice." *Science* 291 (January 19): 424–25.

Lee, Jaeah. 2011. "In Arctic, Warmer Climate Means More Beavers Defecating and Disease." *Mother Jones*, March 4, n.p.

LeMay, Connie. 2013. "No Longer Dead in the Water: Last Year's Ailing Wild Rice Crop Bounces Back." Indian Country Today Media Network, September 23. http://www.Indiancountrytodaymedianetwork.com/2013/09/23/no-longer-dead-water-last-years-ailing-wild-rice-crop-bounces-back-151402

Ljunggren, David. 2000. "Effects of Global Warming Clear in Canada Arctic." Environmental News Network, April 20. http://www.enn.com/enn-subsciber-news-archive/2000/04/04202000/reu_arctwarm_12170.asp

Lynn, Kathy, John Daigle, Jennie Hoffman, Frank Lake, Natalie Michelle, Darren Ranco, Carson Viles, Garrit Voggesser, and Paul Williams. 2013. "The Impacts of Climate Change on Tribal Traditional Foods." *Climatic Change* 120: 545–56.

Maslanik, J. A., M. C. Serreze, and T. Agnew. 1999. "On the Record Reduction in 1998 Western Arctic Sea Ice Cover." *Geophysical Research Letters* 26, no. 13: 1905.

Mathews-Amos, Amy, and Ewann A. Berntson. 1999. "Turning Up the Heat: How Global Warming Threatens Life in the Sea." World Wildlife Fund and Marine Conservation Biology Institute. http://www.worldwildlife.org/news/pubs/wwf_ocean.htm

McGrath, Susan. 2011. "On Thin Ice." *National Geographic*, July, 64–75.

Murphy, Kim. 2001. "Front-Row Exposure to Global Warming; Climate: Engineers Say Alaskan Village Could Be Lost as Sea Encroaches." *Los Angeles Times*, July 8, A-1.

"NASA Study Finds Rising Arctic Storm Activity Sways Sea Ice, Climate." 2008. NASA Earth Observatory, October 6. http://earthobservatory.nasa.gov/Newsroom/NasaNews/2008/2008100627645.html

"Nunavut Capital Battles Heat Wave." 2011. Indian Country Today Media Network, January 5. http://indiancountrytodaymedianetwork.com/article/nunavut-capital-battles-heat-wave-9443

Pegg, J. R. 2004. "The Earth Is Melting, Arctic Native Leader Warns." Environment News Service, September 16.

Pember, Mary Annette. 2012. "Climate Change Threatens the Ojibwe's Wild Rice Harvest." Indian Country Today Media Network, November 2. http://indiancountrytodaymedianetwork.com/article/climate-change-threatens-the-ojibwes-wild-rice-harvest-143579

Qiu, Jane. 2009. "Arctic Ecology: Tundra's Burning." *Nature* 461 (September 3): 34–36.

"Rare Sighting of Wasp North of Arctic Circle Puzzles Residents." Canadian Broadcasting Corporation, September 9. http://www.cbc.ca/story/science/national/2004/09/09/wasp040909.html

Rauber, Paul. N.d. "On Top of the World: Midsummer on Baffin Island: Where the Sun Never Sets on the Polar Bear's Empire and Staying Alive Is a Round-the-Clock Job." http://www.sierraclub.org/sierra/199803/baffin.html

Rolfe, Christopher. 1996. "Comments on the British Columbia Greenhouse Gas Action Plan." West Coast Environmental Law Association presentation to the Air and Water Management Association, April 17. http://www.wcel.org/wcelpub/11026.html

"Scientists Look to 'Traditional Knowledge' to Help Understand Climate Change." 2001. Canadian Broadcasting Corporation News, March 22. http://cbc.ca/cgi-bin/view?/news/2001/03/21/climate010321

Simmonds, Ian, and Kevin Keay. 2009. "Extraordinary September Arctic Sea Ice Reductions and Their Relationships with Storm Behavior over 1979–2008." *Geophysical Research Letters* 36 (October 14): L19715. doi:10.1029/2009GL039810

Sudetic, Chuck. 1999. "As the World Burns." *Rolling Stone*, September 2, 97–106, 129.

Thomas, William. 1998. "Salmon Dying in Hot Waters." Environment News Service, September 22. http://www.econet.apc.org/igc/en/hl/9809244985/hl11.html

"Thunderstorms Are Latest Evidence of Climate Change." 2000. Associated Press Canada, November 15. http://abcnews.go.com/sections/science/DailyNews/arctic_thunder001115.html

"Time to Act on Nunavut Climate Change." 2001. Canadian Broadcasting Corporation, March 16. http://north.cbc.ca/cgi-bin/templates/view.cgi?/news/2001/03/16/16nunmoose

Tingley, Martin P., and Peter Huybers. 2013. "Recent Temperature Extremes at High Northern Latitudes Unprecedented in the Past 600 Years." *Nature* 496 (April 11): 201–5.

Watt-Cloutier, Sheila. 2003. "Speech to Conference of Parties to the United Nations Framework Convention on Climate Change," Milan, Italy, December 10.

Wilkin, Dwane. 1997. "Global Warming Poses Big Threats to Canada's Arctic." *Nunatsiaq News*, November 21. http://www.nunatsiaq.com/archives/back-issues/71121.html#6

Windeyer, Chris. 2011. "South Baffin Swelters in Winter Heat Wave; 'It Doesn't Show Any Signs of Abating.'" *Nunatsiaq News*, January 4. http://www.nunatsiaqonline.ca/stories/article/98789_south_baffin_swelters_in_winter_heat_wave

Wohlforth, Charles. 2004. *The Whale and the Supercomputer: On the Northern Front of Climate Change*. New York: North Point Press/Farrar, Straus and Giroux.

"Yup'ik Villages Ravaged by Fierce Alaska Storms." 2013. Indian Country Media Network, November 20. http://indiancountrytodaymedianetwork.com/2013/11/20/yupik-villages-ravaged-fierce-alaska-storms-152341

# Chapter 5

# *Welcome to My Moonscape*

Native peoples in Alberta have found some of their homelands devastated by tar (or oil) sands mining to the point that the landscape has been compared to a moonscape. Native peoples in the United States also have taken a leading role in opposing the Keystone XL Pipeline, which is being proposed to carry tar sands oil from Alberta to the U.S. Gulf Coast for refining. Roads carrying equipment to the tar sands fields have been blocked, and several arrests have taken place on Nez Perce land in Idaho and the Lakotas' Pine Ridge reservation in South Dakota. Opposition to tar sands development and the Keystone XL has become robust because completion of the pipeline allows tar sands to be further developed, leading to higher greenhouse gas levels in the atmosphere. By 2013, one-third of Alberta's economy was tied in some way to the tar sands (Lizza 2013, 47), including royalties worth more than $4 billion during its 2011–2012 fiscal year.

Canadian Prime Minister Stephen Harper has aggressively promoted tar sands development with plans for 10,000 miles of pipelines eastward and westward across the country as well as the better-known Keystone XL southward from the tar sands fields. By 2014, the Canadian federal Parliament, with Harper's backing, had revoked or annulled 70 environmental laws. In its 2013–2014 budget, the Canadian federal government allocated about $22 million to promote tar sands–based oil outside of Canada (Leslie 2014, A-19). In 2014, Canadian oil producers expected their web of new pipeline capacity to nearly double the country's production of tar sands–based oil by 2025 from 3.5 million to 6 million barrels a day, mostly for export (Krauss and Austen 2014, B-4).

Indigenous opponents of both tar sands mining and the Keystone XL Pipeline have united with non-Native environmental groups to contend that the amount of fossil-fuel energy required to produce useful energy from tar sands exceeds that of conventional oil, contributing to greenhouse gas emissions. Tar sands have the consistency of very gritty peanut butter and are so viscous that they must be thinned (using fossil-fuel energy) to allow refining into usable fuel.

Exploitation of tar sands at the surface requires a form of strip mining that scars the Earth in ways that will not be quickly or easily repaired. The opponents assert that oil sands are a relatively new form of fossil fuel—the last thing the Earth needs when carbon-dioxide levels in the atmosphere have risen to more than 400 parts per million (ppm), more than 40 percent above peak preindustrial levels, with damage to the climate as well as rising seas and oceanic acidity. On a local level, many people worry about oil spills in fragile areas, such as the Nebraska Sand Hills, that could contaminate the Ogallala Aquifer in an area where water is scarce. This aquifer supplies 78 percent of public water and 83 percent of irrigation water in Nebraska, almost a third of the irrigation water used in the United States.

The oil industry and supportive politicians argue that consumers will receive a secure source of vital energy from tar sands. Oil companies produce profits by removing carbon from the Earth and selling it as products that inject carbon dioxide into the atmosphere as a by-product of both manufacture and consumption. Tar sands advocates argue that Alberta collects a carbon tax (it is very small) and that carbon emissions from oil sands mining have dropped about 25 percent since 1990 as refining has become more efficient. Oil sands are still "dirtier" than conventional oil, however, because complex manufacturing is required to make products useful as fuel.

Why is the opposition to tar sands and the Keystone XL ("Express Line" to its supporters or "Extra Leaky" to its opponents) so passionate among both Native peoples and many environmentalists? James Hansen, retired director of NASA's Goddard Institute for Space Studies (who also has been twice arrested at the White House in civil disobedience actions opposing the Keystone XL), says that full use of tar sands means "game over" for a habitable Earth. Hansen joined 19 other scientists in a letter to President Obama on August 3, 2011, saying. "We can say categorically that it's not only not in the national interest [of the United States], it's also not in the planet's best interest" ("NASA Scientist" 2011). A No Tar Sands Caravan to protest the pipeline traversed the United States from California, arriving in Washington, D.C., on August 29, 2011, after a 3,300-mile journey.

According to NASA, the Alberta fields, which were first mined in 1967, are "the world's largest oil sands deposit, with a capacity to produce 174.5 billion barrels of oil—2.5 million barrels of oil per day for 186 years" ("Athabasca Oil Sands" 2011). (The United States as a whole consumes 15 million to 20 million barrels of oil per day.) Environmental activist Bill McKibben has called tar sands mining and the Keystone XL the "fuse to the biggest carbon bomb on the planet" (Tollefson 2013). "Saying that the tar sands are not necessarily worse than coal is like saying that drinking arsenic is not necessarily worse than drinking cyanide," said geophysicist Raymond Pierrehumbert of the University of Chicago. He said that fully developing the tar sands could by itself, even if we suddenly stopped burning coal, "warm the planet an additional 3.6 degrees Fahrenheit by century's end—an amount that climate scientists warn could be catastrophic" (Koch 2014).

The Keystone XL pipeline has become the signature environmental issue of the Obama presidency. Several large U.S. environmental groups called Obama's decision on the Keystone XL a "watershed moment":

> "Dear President Obama," they wrote, "Many of the organizations we head do not engage in civil disobedience; some do. Regardless, speaking as individuals, we want to let you know that there is not an inch of daylight between our policy position on the Keystone Pipeline and those of the very civil protesters being arrested daily outside the White House. This is a terrible project—many of the country's leading climate scientists have explained why in their letter last month to you. It risks many of our national treasures to leaks and spills. And it reduces incentives to make the transition to job—creating clean fuels," the letter states. ("NASA Scientist" 2011)

Signers included the Sierra Club, the Natural Resources Defense Council, Greenpeace, the National Wildlife Federation, Friends of the Earth, the Rainforest Action Network, 350.org., the League of Conservation Voters, Environment America, and the Center for Biological Diversity. Many of these groups (Greenpeace being most active) are also advocating a shutdown of the tar sands fields themselves "and end[ing] the industrialization of a vast area of Indigenous territories, forests and wetlands in northern Alberta. . . . The tar sands are huge deposits of bitumen, a tar-like substance that is turned into oil through complex and energy-intensive processes that cause widespread environmental damage—polluting the Athabasca River, lacing the air with toxins and turning farmland into wasteland," Greenpeace says. "Large areas of the boreal forest are being clear-cut to make way for development in the tar sands, the fastest growing source of greenhouse gas emissions in Canada" ("NASA Scientist" 2011).

## *TAR SANDS AS JUNK ENERGY*

Tar sands are a mixture of clay and sand with bitumen, a thick, low-grade form of petroleum similar to asphalt. According to Thomas Homer-Dixon, who teaches global governance at the Balsillie School of International Affairs, "Tar sands production is one of the world's most environmentally damaging activities. It wrecks vast areas of boreal forest through surface mining and subsurface production. It sucks up huge quantities of water from local rivers, turns it into toxic waste and dumps the contaminated water into tailing ponds that now cover nearly 70 square miles" (Homer-Dixon 2013).

Also, wrote Homer-Dixon (2013), "bitumen is junk energy. A joule, or unit of energy, invested in extracting and processing bitumen returns only four to six joules in the form of crude oil. In contrast, conventional oil production in North America returns about 15 joules. Because almost all of the input energy in tar sands production comes from fossil fuels, the process generates significantly more carbon dioxide than conventional oil production." According to

NASA, oil sands refining produces the equivalent of 86 to 103 kilograms of carbon dioxide for every barrel of crude oil produced. By comparison, 27 to 58 kilograms of carbon dioxide are emitted in the conventional production of a barrel of crude oil.

By 2012, the mining and refining of tar sands in Canada consumed as much natural gas as that country uses for home heating (Kolbert 2007, 49). The gas is used to produce synthetic oil and by-products, such as gasoline. Tar sands require about 15 to 40 percent more energy in manufacture compared to conventional crude oil; oil shales require about twice as much. However, converting tar sands into oil costs about as much as oil, at $30 a barrel. With oil pushing $100 a barrel in late 2014 (but $50 in 2015), tar sands were becoming very profitable for many fossil-fuel companies (Kolbert 2007, 49, 50).

Ryan Lizza (2013, 42), writing in *The New Yorker*, described the mining of tar sands:

> Oil sand has the texture of soft asphalt; twenty percent of it lies close to the surface, and the area is effectively strip-mined. The bitumen-rich sand is removed, mixed with water into a slurry, and spun in centrifuges until the oil is separated, leaving behind vast black tailings ponds that are hazardous to wildlife. The mining operations sprawl ruinously for miles. The remaining eighty percent of the oil sands lie hundreds of feet down beneath a layer of hard rock. Steam is injected deep belowground until the oil naturally separates and is drawn out. The extra energy required to extract the sand makes it a more carbon-intensive fossil fuel—averaging seventeen percent more . . . than conventional oil.

Carol Berry (2012a) of the Indian Country Today Media Network described the environmental damage of tar sands mining:

> If you can imagine the bleak landscape of the moon, you can envision the desolate, 54,000-square-mile tar sands of northern Alberta. . . . "It's literally a toxic wasteland—bare ground and black ponds and lakes—tailings ponds—with an awful smell," said Warner Nazile . . . [an activist] from British Columbia and member of the Wet'suwet'en First Nation. The mining is "despoiling an area roughly the size of England. . . . " University of Alberta scientists "found indications that contamination from the tailings ponds was polluting a huge aquifer that ultimately flows into the Arctic Ocean," Nazile said. Two aboriginal communities downstream from the oil sands have experienced higher-than-average rates of cancer and other health problems, he added.

The Alberta oil sands in 2012 were being mined across an area of 54,900 square miles, according to a satellite data analysis by NASA. About 80 percent of the tar sands lie more than 75 meters below ground, where they are extracted by injecting hot water into a well that liquefies the oil for pumping. The NASA analysis found that pollutants from the mines "are comparable to the emissions from a large power plant or a moderate-sized city," including nitrogen dioxide and sulfur dioxide ("Emissions" 2012). Canada has strict environmental laws requiring the restoration of mine sites. By 2011, one tailings pond had been restored, surrounded by a scarred landscape. The large tailings pond from the first mine, opened in 1967, was drained, filled in, and planted with grass, but NASA satellites didn't show grass growth as of 2011. Two tons of sand are required, along with the water, to produce one barrel of crude oil. Huge trucks remove as much as 720,000 tons of sand per day. The energy required to mine and separate oil from sand adds to its carbon footprint. The mining operation also consumes copious amounts of electric power.

## *TAR SANDS AND CLIMATE CHANGE*

The Keystone pipeline is designed to carry as much as 800,000 barrels of tar sands crude oil daily from to refineries in Oklahoma and Texas from Alberta. It also is routed through the Bakken shale formation of North Dakota and Montana, which is producing new oil from hydraulic fracturing (known as fracking) in which high-pressure chemicals are injected deep underground to break apart the shale rock layer and pump oil out). Fracking also brings up natural gas that, not withstanding its environmental costs, has played a major role in reducing U.S. greenhouse gas emissions by 10 percent between 2005 and 2013 as it replaces dirtier coal in electricity generation (Gillis and Fountain 2014, A-12). Tar sands and fracking for oil tap relatively new sources of fossil fuels that will increase in years to come, raising greenhouse gas levels in the atmosphere. "If Canada proceeds [with oil sands development], and we do nothing, it will be game over for the climate," James Hansen, author of *Storms of My Grandchildren,* wrote in the *NewYork Times*:

> Canada's tar sands, deposits of sand saturated with bitumen, contain twice the amount of carbon dioxide emitted by global oil use in our entire history. If we were to fully exploit this new oil source, and continue to burn our conventional oil, gas and coal supplies, concentrations of carbon dioxide in the atmosphere eventually would reach levels higher than in the Pliocene era, more than 2.5 million years ago, when sea level was at least 50 feet higher than it is now. That level of heat-trapping gases would assure that the disintegration of the ice sheets would accelerate out of control. Sea levels would rise and destroy coastal cities. Global temperatures would become intolerable. Twenty to 50 percent of the planet's species would be driven to extinction. Civilization would be at risk. . . . If this sounds apocalyptic, it is. (Hansen 2012)

The concentration of carbon dioxide in the atmosphere has risen from 280 ppm to slightly more than 400 ppm over the past 150 years, as of 2014. The tar sands contain enough carbon—240 gigatons—to add 120 ppm. Tar shale, similar to tar sands found mainly in the United States, contains at least an additional 300 gigatons of carbon. If we turn to these dirtiest of fuels, instead of finding ways to phase out our addiction to fossil fuels, there is no hope of keeping carbon concentrations below 500 ppm—a level that would, as Earth's history shows, leave our children a climate system that is out of their control.

In an attempt to address critics of oil sands carbon emissions, Alberta's government has invested in a $1.35 billion demonstration project to capture and bury as much as 1.2 million tons of carbon dioxide a year from an oil sands facility that manufactures oil from bitumen. Alberta also has required reductions in carbon intensity (emissions per unit of production) by 12 percent, or payment of a $14-per-pound carbon tax. Between 2007 and 2012, that tax raised $378 million for clean energy investments, of which about $80 million was spent on projects related to oil sands (Tollefson 2013).

### *TAR SANDS OIL SPILLS*

Opponents of the Keystone XL pointed to the less-than-sterling record of existing pipelines, especially recent leaks in communities that were bearing high cleanup costs, such as an Enbridge Energy spill near Marshall, Michigan, in 2010, spilling more than 840,000 gallons of tar sands crude oil. In March 2012, an Exxon pipeline sprang leaks in Mayflower, Arkansas, a town of 2,200 near Little Rock, provoking the evacuation of two dozen homes. This pipeline also was carrying tar sands crude oil, of which about 210,000 gallons spilled.

Both pipeline companies have spent tens of millions of dollars trying to recover the heavy crude, similar to the product that Keystone XL would carry. River and floodplain ecosystems have had to be restored, and neighborhoods are still being refurbished. Legal battles are being waged, and residents' lives have been forever changed. "All oil spills are pretty ugly and not easy to clean up," said Stephen K. Hamilton, a professor of aquatic ecology at Michigan State University who advised the Environmental Protection Agency (EPA) and the state on the cleanup in Marshall. "But this kind of an oil is even harder to clean up because of its tendency to stick to surfaces and its tendency to become submerged" (Frosch 2013). A 40-mile segment of the river that traverses Marshall, the site of the Enbridge, was closed for two years, and several waterfront homes were abandoned. In the meantime, Enbridge underestimated the amount of oil that had spilled by a factor of 100. The EPA ordered the company to dredge the river. Enbridge also bought out 154 homes in the area most intensely affected by the spill and spent several million dollars to enhance or build roads and parks along the oil-soaked river.

In Mayfield, four months later, according to an account in the *New York Times*, "the neighborhood of low-slung brick homes is largely deserted, a ghostly column of empty driveways and darkened windows, the silence broken only by the groan of heavy machinery pawing at the ground as remediation continues." Exxon offered to buy the 22 vacated residences but only at their postspill value. The company also had spent $2 million on temporary housing for residents and more than $44 million on the cleanup by mid-2013 (Frosch 2013).

## *ON THE GROUND IN TAR SANDS COUNTRY*

In the midst of the debate over the environmental effects of tar sands, in late August 2013 a large toxic waste spill, the largest of its kind in North America, rolled over indigenous lands in northern Alberta. The *Toronto Globe and Mail* reported that "the substance is the inky black colour of oil, and the treetops are brown. . . . Across a broad expanse . . . the landscape is dead. It has been poisoned by a huge spill of 9.5 million litres of toxic waste from an oil and gas operation in northern Alberta, the third major leak in a region whose residents are now questioning whether enough is being done to maintain aging energy infrastructure" (" 'Every Tree' " 2013).

"Every plant and tree died," Dene Tha' First Nation Chief James Ahnassay told the *Toronto Globe and Mail* (" 'Every Tree' " 2013). The Dene Tha' said that the spill ran along a trapline (animal-harvesting area) about half a mile from its reserve, a mile from the same people's fishing grounds in the Zama River, where its members fish. The 103-acre spill sprang from a breach in a pipeline owned by the Apache Corporation near Alberta's far northern border. Company representatives said that it was salty water tainted with a small amount of oil and that no people were harmed.

Such a spill is not a one-time event. A 1,000-person Beaver Lake indigenous community near Lac La Biche (in northern Alberta) now lives with a moonscape all the time and has been nearly surrounded by oil sands extraction, as its hunting and fishing range has been laced by 600 miles of roads. In 2008, the Beaver Lake Cree Nation filed suit against the governments of Alberta and Canada alleging breach of treaty rights. The defendants fought to have the suit dismissed (as "frivolous") but failed as the Court of the Queen's Bench upheld the community's standing in court in a ruling by Justice Beverley Browne. The governments appealed, and on April 30, 2013, the Alberta Court of Appeal again upheld their right to sue. The suit claims comanagement rights under the treaty.

Brandi Morin of the Indian Country Today Media Network quoted Beaver Lake Cree Nation attorney Drew Mildon as saying that should this case be successful, it could be a precedent-setting impediment to oil sands development, with repercussions throughout the area. "This is the thing that can keep tar sands development from moving forward," he said (Morin 2013). Mildon said that the effects of oil production through steam-assisted gravity drainage, which is most common in the area, is subtly damaging. "You end up with oil in the water table, with animals being poisoned, with forests being so fractured that it no longer supports any of the fur bearers or other animals that First Nations depend on for food" (Morin 2013). "Science fiction is rife with fantasies of terra-forming—humans traveling to lifeless planets and engineering them into earthlike habitats," wrote Naomi Klein (2014). "The Canadian tar sands are the opposite: terra-deforming. Taking a habitable ecosystem filled with life and engineering it into a moonscape where almost nothing can live" (Klein 2014, 139).

British Columbia photographer Garth Lenz said, "I'd heard about the tar sands but I hadn't been, so I went there and spent a couple of days and was pretty much flabbergasted by the scale of the devastation and the impacts. I had photographed industrial devastation all over, including some of the most massive clear-cuts on the planet, right in British Columbia and in Chile and Patagonia, so I'd seen that massive industrialization of the landscape on a huge, huge scale," he said. "But I was completely unprepared for what I found. Because this is just completely off-the-grid crazy—the scale is unbelievable" ("Athabasca Oil Sands" 2011).

Lenz pointed out that the development harms fish and caribou that indigenous peoples in the area rely on. "It's the complete eradication of an ecosystem," Lenz said. "I mean, the forest is clear-cut, the wetlands are drained and dredged, the soil is dug up, replaced by massive mines and toxic ponds which you can see from outer space" ("Athabasca Oil Sands" 2011). Chief Bill Erasmus of the Yellowknife Northwest Territories in northern Canada said that "our people, in some areas, can no longer eat the fish. . . . Our people can no longer drink the water. Water levels are decreasing. Where I'm from, it's never been like that before" (Capriccioso 2011).

### *TRIBES AND NATIONS FORGE ALLIANCES*

In August 2011, the National Congress of American Indians (NCAI) declared opposition to the Keystone XL Pipeline. "Based on the relatively poor environmental record of the first Keystone pipeline, which includes numerous spills, U.S. regulators shut the pipeline down in late May, 2011," said the NCAI resolution, which concluded that "it is probable that further environmental disasters will occur in Indian country if the new pipeline is allowed to be constructed." The Assembly of First Nations of Canada also has expressed similar opposition. The NCAI urged the United States to reduce dependence on oil from tar sands that makes such a pipeline necessary with "work towards cleaner, sustainable energy sources" ("NCAI Condemns" 2011). "Homeland and economic security starts with energy security, but Indian Country wants it to be done right," said NCAI President Jefferson Keel, "not at the expense of the health of our communities and resources, both tribal and non-tribal. During challenging economic times in our country and in our tribal nations, domestic energy when developed responsibly can create jobs while ensuring that our people and natural resources remain safe and plentiful" ("NCAI Condemns" 2011).

People from roughly 25 tribes and nations in Canada and the United States met on Yankton Sioux (Ihanktonwan) lands for three days in late January 2013 to draft a common statement opposing Canadian tar sands and the Keystone XL Pipeline. They called the meeting the Gathering to Protect the Sacred from the Tar Sands and Keystone XL. A signing ceremony was held, along with honoring songs and statements by Chief Arvol Looking Horse, Keeper of the White Buffalo Calf Pipe Bundle and a spiritual leader of the Lakota, Dakota, and Nakota, as well as other notable people, including First Nation actress Tantoo Cardinal (who is Cree) and Debra White Plume (an Oglala Lakota) of the activist group Owe Aku (Bring Back the Way). White Plume and Cardinal had been arrested in front of the White House in 2011 protesting Keystone XL. White

Plume also was arrested in 2012 on Pine Ridge along with others who blockaded trucks carrying loads of machinery headed for the Alberta oil sands mines.

"We're here to stand together to protect Mother Earth [for] future generations," Looking Horse said at the treaty meeting. "To us, treaties are the supreme law. We came together with good hearts, good minds." The common statement says that oil sands mining presents "unacceptable risks to the soil, the waters, the air, sacred sites, and our ways of life" (Berry 2013). The alliance's statement asserts that pipelines "carry the possibility of pipeline and tanker oil spills, pose health and ecological threats, and infringe on sacred and historic places, burial grounds and irreplaceable cultural resources" (Berry 2013).

This alliance was forged after an initial meeting in late 2012 when about 20 spiritual leaders of indigenous peoples across North America gathered at a sun dance in South Dakota to compose a declaration "to protect Mother Earth from the impacts of development in Canada's oil sands mega-project." Mohawk, Lakota, Navajo, Apache, Aztec, and Ojibwe leaders took part, hosted by Sicangu Leonard Crow Dog, Lakota spiritual leader. Sun dancers who had come from an area spanning Alaska to Peru pledged their support as well. The idea for the statement originated with Sundance Chief Rueben George of the Tsleil Waututh Nation in North Vancouver, British Columbia. Crow Dog, who has been an American Indian spiritual adviser since the 1970s, said, "It is our responsibility to protect and care for these elements in accordance with our own sacred laws and traditions, and in doing so, maintain our spiritual relationship to the land, water, plants, animals, our ancestors, all of our relations and future generations" (Ball 2012b).

A declaration by the spiritual leaders said, in part, "We therefore will defend our land and exercise our own laws and traditions from all directions to oppose in all tar sands pipeline and tanker projects, and the tar sands development itself, all of which threaten the physical, mental, emotional and spiritual well-being of all of our relations" (Ball 2012b). "The Keystone pipeline goes right through their [Lakota] territory," George told the Indian Country Today Media Network. "It was beautiful to have so many chiefs come together, because [be it] in Peru or up in Alaska or here in Vancouver, we're dealing with the same problems: fossil fuel problems. We also spoke of alternatives, too. Our ceremonies and beliefs come from the elements of Earth: from fire, earth, water and sky. That's why we believe our lands and waters are sacred" (Ball 2012b).

On October 1, shortly before the spiritual leaders issued their statement, the Athabasca Chipewyan First Nation (ACFN), whose lands are situated near the oil sands mines, issued a constitutional challenge based on Treaty 8 asserting that Alberta's government and Shell Canada had violated their legal rights by failing to consult them regarding the development of the mines. The Chipewyan demanded "mandatory consultation with First Nation treaty signatories. It also demands that their right to access their territories for hunting, trapping, harvesting and other traditional uses be honored. The Chipewyan alleged that the Jackpine expansion, by damaging the environment, would prevent them from engaging in those activities and resources and thus violate the treaty"

(Ball 2012a). Treaty 8 guarantees signatories a right to self-governance and use of their lands and resources for sustenance. ACFN spokesperson Eriel Deranger said,

> What are the costs of pushing the industry through? We're talking about doubling production in the tar sands. We're already having problems with the current pace of development. Doubling it is psychotic. Some people think the tar sands and First Nations people can coexist, [but] I don't know how you could possibly rip up thousands of kilometers of boreal forest and traditional territories, de-water, poison and contaminate river systems, and consider that a plausible way for coexistence? (Ball 2012a)

In 2014, a report by the Joint Oil Sands Monitoring Program alleged that dams holding back wastewater from oil sands mining waste were seeping 6.5 million liters (or 1.7 million gallons) daily into the Athabasca River's groundwater, including bitumen, arsenic, cadmium cyanide, phenols, naphthenic acids, and other metals. "In short, [the study] highlights past studies identifying tailings ponds as significant sources of groundwater contamination, and brings to light that groundwater contaminated by leaking tailings ponds is almost certainly flowing into the Athabasca River," said William Donahue, a freshwater science specialist based in Edmonton (Nikiforuk 2014).

Shell Canada was proposing to increase oil sands production by 100,000 barrels a day as part of the Jackpine project in a 31,429-square-mile "disturbance area" that indigenous peoples use for hunting, trapping, and fishing in the Muskeg River area. Shell spokesperson David Williams said that the company had consulted with the First Nation repeatedly and that it does respect ACFN's treaty rights. "Shell has engaged extensively with the Athabasca Chipewyan for more than 15 years," he told the *Toronto Globe and Mail*. Deranger said that mere contact did not constitute consultation and that meeting solely with Native people who stood to benefit from the mining did not meet the terms of the treaty.

"There are First Nations who think the tar sands are great," Deranger admitted. "People have jobs. People now can afford to take their kids to Edmonton to go to the dentist. These are luxuries for people. But we have to start weighing the costs . . . [including] hundreds of toxic tailings ponds, open pit mines, significant emissions and polluted rivers across a giant swathe of Alberta," Deranger said (Ball 2012a). Richard Ray Whitman of the Yuchi/Muskogee Creek tribes said that many American Indians in the United States support Canadian First Nations that have been affected by pipeline development. "Our concerns also go far beyond what might happen here," he said. "Because of the extraction of the tar sands, we are currently witnessing the devastation of lands considered sacred by indigenous people in Canada. Opposing the Keystone XL Pipeline means standing in solidarity with all our Native brothers and sisters in the Northern U.S. and Canada" (Capriccioso 2013).

*NATIVE PEOPLES UNITE AGAINST THE KEYSTONE XL*

Delegations of Native peoples have visited the State Department and the White House as well as the EPA, expressing their opposition to the Keystone XL and other pipelines that are planned to cross over or that are close to their lands. In May 2013, 10 U.S. Native tribes and nations in the pipeline's proposed line of transit declared their opposition:

> On this historic day of May 16, 2013, ten sovereign Indigenous nations maintain that the proposed TransCanada/Keystone XL pipeline does not serve the national interest and in fact would be detrimental not only to the collected sovereigns but all future generations on planet earth. This morning the following sovereigns informed the Department of State Tribal Consultation effort at the Hilton Garden Inn in Rapid City, SD, that the gathering was not recognized as a valid consultation on a "nation to nation" level. ("Full Text" 2013)

The statement was signed by the Southern Ponca, Pawnee, Nez Perce Nation, Oceti Sakowin (Seven Council Fires People), Sisseton-Wahpeton Oyate, Ihanktonwan Dakota (Yankton Sioux), Rosebud Sioux, Oglala Sioux, Standing Rock, Lower Brule Sioux, Cheyenne River Sioux, and Crow Creek Sioux. At about the same time, the Great Plains Tribal Chairman's Association (which includes 16 tribal presidents in Nebraska, South Dakota, and North Dakota) also supported this position, which expresses concern regarding "potential pipeline impacts on natural resources, especially our water: potential spills and leaks, groundwater and surface water contamination," as well as climate change made worse by the exploitation of fossil fuels. "America would be better served," the common statement said, "by a comprehensive program to reduce its reliance on oil, and to invest in the development and deployment of sustainable energy technologies, such as electric vehicles that are charged using solar and wind power." The statement also opposed the pipeline on treaty-rights grounds: "If the Keystone XL pipeline is allowed to be built, TransCanada, a Canadian corporation, would be occupying sacred treaty lands as reserved in the 1851 and 1868 Fort Laramie Treaties. It will be stopped by unified resistance" ("Full Text" 2013).

On May 16, 2013, elders and chiefs of at least 10 Native American nations walked out of a meeting with U.S. State Department officials in Rapid City, South Dakota, on May 16 in which the government was attempting to engage in tribal consultation over the Keystone XL Pipeline. Many in the group that earlier had signed a joint statement opposing the Keystone XL (Southern Ponca of Oklahoma, Pawnee Nation, Nez Perce Nation, Sisseton-Wahpeton Oyate, Ihanktonwan Dakota Yankton Sioux, Rosebud Sioux, Oglala Sioux, Standing Rock, Cheyenne River Sioux, the and Crow Creek Sioux) called the meeting "invalid" and asked to meet directly with President Obama on pipeline-related issues.

The Native leaders who walked out of the meeting with State Department officials issued a statement objecting to the pipeline for several reasons. "On this historic day of May 16, 2013, ten sovereign Indigenous nations maintain that the proposed TransCanada/Keystone XL pipeline does not serve the national interest and in fact would be detrimental not only to the collected sovereigns but all future generations on planet Earth. This morning the following sovereigns informed the Department of State Tribal Consultation effort at the Hilton Garden Inn in Rapid City, SD, that the gathering was not recognized as a valid consultation on a 'nation to nation' level," they said. "Eventually all remaining tribal representatives and Tribal Historic Preservation Officers left the meeting at the direct urging of the grassroots organization Owe Aku," the chiefs said in their statement. "Owe Aku, Moccasins on the Ground, and Protect the Sacred are preparing communities to resist the Keystone XL pipeline through Keystone Blockade Training" ("Chiefs" 2013).

The NCAI kept detailed watch on the State Department's handling of the Keystone XL issue, severely criticizing sections of its report on oil-spill prevention and remediation as well as its lack of detail on American Indians' and Alaska Natives' water quality and supply. "In total," the NCAI statement concluded, "if these concerns are not addressed sufficiently or mitigated to the fullest extent, it is in the best interest of the United States to reject the Keystone XL pipeline permit solely on the basis of the federal trust responsibility to tribal nations. . . . The project as outlined in the DSEIS poses tremendous risks to the cultural and natural resources of tribal nations and is not in the best interest of the tribal nations and their citizens" ("Fill Gaps" 2013).

## *A BLOCKADE AT PINE RIDGE*

The Oglala Sioux on the Pine Ridge Indian reservation expressed concern that the path of the Keystone XL would cross a pipeline delivering water to them as well as the neighboring Rosebud Sioux in two places (Woodard 2013). A blockade at Pine Ridge began on March 5, 2012, after a resident of Wanblee on the reservation was forced off the road by a convoy of 150-ton semitrailers. She called several neighbors, who converged on the site, blockading the road with their cars and stopping the TOTRAN Transportation Services rigs. Traffic stopped for six hours as about 40 cars joined the blockade, containing between 50 and 75 people. Tribal police arrested five people who refused to move (Alex White Plume, Debra White Plume, Sam Long Black Cat, Andrew Iron Shell, and Don Iron Shell) on charges of disorderly conduct.

The trucks' drivers said they were destined for Canadian mines, hauling "treater vessels," which use intense heat to separate oil and gas from other substances, invoiced at almost $260,000 per load. The *Rapid City Journal* reported the next day that TransCanada denied that the shipments were part of Keystone XL construction. However, Oglala Nation Vice President Tom Poor Bear said that South Dakota Governor Dennis Daugaard's secretary of tribal relations, J. R. LaPlante, told him that the trucks had been routed through Pine Ridge by TransCanada to avoid paying $50,000 per vehicle in fees for using South Dakota highways. "He said, 'Mr. Poor Bear, I want to apologize. The South Dakota

Department of Transportation'—and then he named a couple senators and himself—'had a meeting a couple weeks ago to reroute these trucks that are holding these pipes and water tanks that are going to Canada for the Keystone pipeline. . . . We had to reroute them through your reservation.' "

Poor Bear then replied, " 'Mr. LaPlante; I took a lead in opposing the Keystone pipeline, as well as our neighbors the Rosebud Sioux Tribe, and said that we oppose the pipeline and you allowed them to come through our reservation without asking permission?' " (Schilling 2012). Poor Bear had called out to President Obama as he spoke to university students in Denver during October 2012 urging him to disapprove a permit for the $7 billion, 1,711-mile Keystone XL Pipeline. The president replied, "I hear you," and told the delegation that "no decision has been made" (Berry 2013). Speaking for an Oglala Sioux delegation, Poor Bear said, "The fight against this pipeline is far from over; we must become as one to protect our mother, the earth, and future generations," urging "on behalf of the Oglala Lakota" for "each and every one of our relatives from other tribes and nations to stand with us in unity and solidarity to protect what is sacred." The Oglala Sioux Tribal Council also passed a resolution against the pipeline, saying that it would involve "accessing a 300-foot-wide corridor through unceded treaty lands of the Great Sioux Nation" as included in the Fort Laramie treaties of 1851 and 1868 (Berry 2013).

## *BLOCKADE BY THE NEZ PERCE*

The Nez Perce hosted Lewis and Clark in 1805 and later, in 1877, led the U.S. Army on a months-long chase over some of the roughest landscape in North America. In 2013, the fighting spirit of the Nez Perce again was marshaled by transport through their lands of huge loads destined for Alberta's tar sands fields. On August 5, 2013, more than 200 Nez Perce intercepted a convoy of trucks destined for the oil sands fields of Alberta, forming a blockade that was broken up by police, who arrested 30 people, including chairman Silas Whitman and six members of the tribe's executive council.

In 2013, as in 1877, the Nez Perce didn't appreciate being pushed around, and they quickly put events into context. "The development of American corporate society has always been—and it's true throughout the world—on the backs of those who are oppressed, repressed or depressed," Silas Whitman, chairman of the Tribal Executive Committee, told the *New York Times*. "We couldn't turn the cheek anymore." After their meeting, the Nez Perce leaders decided to face arrest as a group (Johnson 2013). Whitman, 72 years of age at the time, was one of several Nez Perce arrested at the barricade as police dismantled it. The blockade lasted four nights.

"Lights! Lights!" they shouted as amber-colored flashers on a pilot truck came into view in the darkness just after midnight along Highway 12 on the Nez Perce reservation. The Indian Country Media Network reported that "a rush of more than 200 tribal members and others—from grandmothers to children—followed, all determined to stop a football-field-sized mega-load from passing through their sacred lands," erecting wooden barriers, as "police cruisers idling in the darkness nearby flipped on their headlights and rooftop light bars and rolled to within yards of the blockade. It was an eerie blue-and-red strobe-lit standoff: Tribal members sang and drummed and whooped, while state and tribal police faced them, arms crossed. Looming behind the police came the mega-load, its cylindrical face appearing as an enormous ghostly moon, swathed in a white tarp" (Taylor 2013). The trucks they stopped were carrying a 23-foot-tall, 322-ton water evaporation hooked onto a diesel truck front and back, 243 feet long.

Early in 2013, the Nez Perce Executive Council adopted a position opposing transportation of very large (megaload) mining equipment through its land along 780 miles of Highway 12. Such shipments have become common there. "They will also be traveling really close to our creation story," said McCoy Oatman, chairman of the Nez Perce Executive Committee. "It's called 'The Heart of the Monster.' It's essentially the birthplace of the Nez Perce people. It [the route] runs dangerously close to that site, and that's pretty significant for us" ("Nez Perce Victory" 2013).

The protests paid off. The Nez Perce also joined with Idaho Rivers United, an environmental group, and took the matter to Boise's federal district court, where Chief Judge B. Lynn Winmill stopped further transport of megaloads until the Nez Perce and the U.S. Forest Service assessed their environmental impact. The Nez Perce believe that their scenic route was being turned into an industrial corridor as federal and state officials ignored them. In September 2013, the Forest Service closed the Nez Perce route to megaloads and required consultation with the tribe in the future. "The closure order is effective immediately and is in place until the agency agrees to lift it, mostly likely when forest officials complete a study on the impacts the shipments could have on the river corridor," the Associated Press said. "The agency has also vowed to consult with tribal officials on the impact to treaty rights and cultural values" ("Nez Perce Victory" 2013).

## *PROTESTS SPREAD*

Protests of the megaloads soon spread beyond the Lakota and Nez Perce. The Ho-Chunk (Winnebago) Tribal Council adopted a resolution against all sand mining on their lands in 2012. The sand was to be used for fracking. Such operations have been spreading across North America, and Native Americans have taken a leading role in resisting some of them. Water is used in very large amounts at high pressure to fracture shales, releasing the gas, in combination with a "fracking cocktail [that] includes acids, detergents and poisons that are not regulated by federal laws but can be problematic if they seep into drinking water" (Brantley and Meyendorff 2013). Methane release can create the risk of explosion.

On August 9, 2013, about 150 people, Native and non-Native, arrived at Pier 96 in New York City, having paddled as far as 300 miles along the Hudson River and its tributaries, in defense of the Two Row Wampum, a mutual respect treaty first negotiated in the early seventeenth century between the Haudenosaunee (Iroquois) and immigrating Europeans. They docked their canoes and kayaks, then hiked to the UN headquarters building on the eastern side of Manhattan Island. Their purpose was both historical and environmental, and one of their major targets was fracking, a call that was renewed on January 8, 2014, at Governor Andrew Cuomo's State of the State Address in Albany ("Call to Action" 2014).

Riders undertook a 150-mile horseback journey from the Pine Ridge reservation to the Cheyenne River reservation, both in South Dakota, near the route of the proposed Keystone XL Pipeline. "Thundering across the plains on horseback, along the routes of two proposed oil pipelines, Earth's Army has wound up its journey to draw attention to not just TransCanada's Keystone XL pipeline, but also a lesser-known one being proposed by Enbridge across White Earth territory," said one eyewitness account. The ride was led by Percy White Plume, whose ancestors survived the 1890 Wounded Knee massacre. "We can drink bottled water, but our relatives in the horse nation, the buffalo nation and the animals cannot drink bottled water, our water is sacred," he said ("Anishinaabe" 2013). The riders were organized by the Horse Spirit Society of Wounded Knee and supported by the Swift Family Foundation, 350.org , the U.S. Climate Action Network, and Honor the Earth.

At about the same time, nine people were arrested on June 24, 2013, after they chained themselves to a work trailer and an excavator in Seminole, Oklahoma, in protest of the Keystone XL, blocking construction of a pump station for the pipeline on Seminole treaty land. Halfway across North America, the Columbia River Inter-Tribal Fish Commission joined the opposition. "The development, transportation, and use of Canada tar sands oil will have long-lasting negative effects and pose significant threats to the Columbia River Basin, its natural resources and the people who reside there," stated a resolution passed on August 23, 2013, by the Umatilla, Yakama, Warm Springs, and the Nez Perce, who had been protesting the megaload transports through their lands for three years by 2013.

Native people at Umatilla and the Confederated Tribes of Warm Springs in northern Oregon also protested the passage through their lands of a 45-ton megaload (18 feet tall, 22 feet wide, and 376 feet long) bound for Alberta's tar sands fields in early December 2013. About 70 people carrying signs reading "Hands Off Our Planet! Stop the Tar Sands" swarmed the megaload and impeded its path. "This has gone too far. Our children are going to die from this," some of them said in an online video posted by Portland Rising Tide. "If we don't stop this now . . . [future generations] are going to ask, 'Where were you? Where was our tribe?' " (Dadigan 2013). A few locked their bodies to the huge truck. The megaload forced the protesters aside but was later stopped by snow and ice near Pendleton, Oregon. Several activists were arrested on charges of disorderly conduct after police removed the locks.

Alberta's Lubicon Lake Cree Nation protested fracking as well as tar sands mining on and near their lands, the latter with a blockade on unceded territory of a road northeast of Peace River in late 2013. The Calgary-based oil and gas company Penn West Petroleum Ltd, whose site was being dismantled, filed suit to dismantle the Lubicons' blockade. "The judge denied [us] the opportunity to raise any of the constitutional issues and arguments for the Lubicon," said Garrett Tomlinson with the Lubicon Lake Nation. "More time must be provided for both sides to be heard" (Troian 2014). Some of the Lubicon argued that lack of a treaty between their tribe and the Canadian government rendered any permits to Penn West null and void. The Lubicon Cree leadership itself is split, however, between a faction supporting Penn West and one opposing it.

The Lubicon Lake Nation's leadership argued that Canada has never entered into a treaty with them. They assert that lack of a treaty renders permits for oil and gas development on Lubicon lands issued to Penn West by Alberta null and void. Penn West has explored fracking between Haig and Sawn lakes, areas where Lubicon Lake people are known to carry out traditional activities such as fishing and hunting. The Indian Country Today Media Network reported that a "dialogue among the Lubicon themselves may be difficult, since the people are actually divided into two bands. Chief Bernard Ominayak represents the Lubicon Lake Nation, whereas Chief Billy Joe Laboucan leads the Lubicon Band, a separate group that was federally recognized by Aboriginal Affairs and Northern Development Canada in February 2013. . . . The First Nation split during the mid-1990s, a breakup that some say was orchestrated by Aboriginal Affairs in order to undermine an unfinished land claim settlement with Chief Ominayak's group. Chief Ominayak and his council said that by supporting industry, Chief Laboucan's group is out to destroy the Lubicon lands, as are PennWest and the provincial government" (Troian 2014).

"We have to stand up for Mother Earth. We have to stand up for our sacred water—for our children, our grandchildren, for the coming generations," said Lakota activist Debra White Plume as she was arrested at the White House on September 2, 2011, with 186 others protesting the Keystone XL Pipeline. Several of those arrested were Canadian Native people who also were protesting oil sands development on or near their homelands. White Plume said that leaks from the pipeline could desecrate the freshwater Ogallala Aquifer near her homeland in Pine Ridge, South Dakota, a violation of the 1868 Fort Laramie Treaty. "It is with great honor that I come here today to ask President Obama to stand with us for Mother Earth against Father Greed," Plume said (Capriccioso 2011).

On April 22, 2014, cowboys and Indians united in opposition to the Keystone XL on the Mall in Washington, D.C., many of them on horseback, to the sound of beating drums, as about 100 people gathered around a large tipi. The demonstration was assembled by the Cowboy and Indian Alliance, representing landowners and Native opponents of the pipeline who marched and rode along Independence Avenue onto the Mall.

When 68-year-old Canadian folksinger Neil Young devoted a tour to help the ACFN's legal campaign to impede expansion of oil sands mining, Prime Minister Stephen Harper said that his "rock-star lifestyle made him unfit" to criticize it. In response, Young took the stage at a concert in Toronto with an audience of 2,700 on January 12, 2014, with song lyrics needling Harper as a party to genocide. David Ball (2014a) of the Indian Country Today Media Network wrote that "the legendary Canadian musician substituted Harper's name into his song

'Pocahontas,' which describes massacres against Indigenous Peoples." The Chipewyan First Nation says that contamination of soil and water from tar sands mining is causing cancer rates among their people to rise rapidly. "In the street outside Massey Hall," wrote Bell, "dozens held a round dance—a fixture of the Idle No More movement that has swept the country over the past year."

In addition to Toronto, Young's "Honor the Treaties" tour visited Winnipeg, Regina, and Calgary. "We made a deal with these people," Young said. "We are breaking our promise. We are killing these people. The blood of these people will be on modern Canada's hands, and it will be the result of not just a slow thing, but of a fast and horrific thing if this continues. Believe me, these people are not going to sit back and let modern Canada roll over them" (Ball 2014a).

A celebrity filmmaker arrived in tar sands country in Fort Chipewyan, Alberta, and met with members of the ACFN to discuss creating a documentary on the Alberta tar sands. On August 22, 2014, actor Leonardo DiCaprio met with Black Swan director Darren Aronofsky, both of whom received a blast of invective from a spokesman for the Canadian Association of Petroleum Producers, wrapped in Canadianismo: "Like Canadians, we are growing tired of the fad of celebrity environmentalists coming into the region for a few hours or a few days, and offering their ideas and solutions to developing this resource" ("Leo DiCaprio" 2014).

President Obama was met with protests of the XL Pipeline when he became only the fourth sitting U.S. president to visit an Indian reservation on June 13, 2014. "President Obama must reject this pipeline and protect our sacred land and water . . . [it] is a death warrant for our people," said Bryan Brewer, president of the Oglala Sioux tribe, as the Indigenous Environmental Network demonstrated against the project. Obama's remarks at a powwow on the Standing Rock Sioux reservation in central South Dakota focused on increasing economic opportunities for Native Americans.

In early 2015, the Fort Peck reservation's Sioux and Assiniboine peoples of Montana passed a resolution opposing the Keystone XL, saying that "the pipeline would jeopardize drinking water projects for the reservation." The pipeline, as proposed, would cross watercourses upstream from the Fort Peck Dam, sites from which a rupture could contaminate the entire Missouri River water supply. This route includes the intake for the Assiniboine/Sioux Water Project on the Fort Peck reservation. "This project will provide water for at least 30,000 people," said Bill Whitehead, a tribal member, former councilman, and former state legislator who is now chairman of the Assiniboine/Sioux Water Project. "Through our treaty rights and water compacts it covers all of northeastern Montana, including the reservation and four counties" (McNeel 2015).

Native Americans, the largest minority group in South Dakota at 9 percent of the population, have been vocal opponents of the pipeline; 313 miles of its proposed route runs through the state. Some Rosebud Sioux had maintained a "spirit camp" opposing the pipeline for more than a year by mid-2015 (with a trailer and a tent and without running water) on tribal land near the Keystone XL route, where, according to a report in the *New York Times*, they have "vowed to use the site as a base camp for protesters if construction ever begins.

Several activists, including Mr. Frederick, have said they would risk arrest through civil disobedience" (Smith 2015). Faith Spotted Eagle, the chairwoman of the Yankton Sioux tribe's Treaty Council, has emerged as a leader of an increasingly organized coalition of Native Americans, landowners, and grassroots groups seeking to block construction of the Keystone XL across South Dakota.

Invoking Article I of the 1868 Fort Laramie Treaty, the Lower Brule Sioux tribe voted in April 2015 to reject the Keystone XL Pipeline and evict TransCanada from its lands "in direct response to the unethical business practices that TransCanada has demonstrated over the last six years." Acting Chairman Kevin Wright said that "as descendants of the people of this land we have witnessed destruction of many magnitudes, we are concerned for our land, water, and most importantly not only the physical wellbeing of our people but spiritual wellbeing as well. I am first a human being, not a politician, when it comes to these matters. I believe in protecting our people and look to more ecological ways of living" ("Lower Brule" 2015). The treaty allows the eviction of "bad men" who do not act in good faith. Sioux land under the 1868 treaty included about 40 percent of present-day South Dakota, including part of the Keystone XL's proposed route.

Following a commencement speech in May 2015 at the Lake Area Technical Institute in Watertown, South Dakota, President Obama accepted a star quilt created by Sisseton-Wahpeton Dakota artist DeVon Burshiem, who had embroidered "NOKXL" ("No Keystone XL") on its reverse side. "*Tasina Pezuta Win emakiyapi ye* (They call me medicine blanket woman). It was truly an honor and a blessing to know that the quilt I was making would soon be in the hands of the President of the United States, Barack Obama!" said Burshiem by way of the Indigenous Environmental Network. "Heartfelt prayers were made during that time, tears ran down my face as I prayed for our land and for our water, *mni wicozani* [water is life], without water we are without life. #NOKXL" ("Message" 2015).

## *SHIPPING OIL—AND SPILLING IT—BY RAIL*

Lacking approval for the Keystone XL and other pipelines, oil companies have taken to shipping their product by train, 100 tank cars or more at a time, leading in some cases to spectacular wrecks and spills (such as in North Dakota and Quebec in 2013). More than 400,000 tank cars of crude were shipped in 2013, up from 10,000 in 2008. Two-thirds of the Bakken formation's shale oil departed by train in 2013. At the same time, the number of gallons of oil spilled by railroad in 2013 exceeded that of the previous 37 years (1975–2012) combined, according to the Association of American Railroads (Krauss and Mouawad 2014, A-1). Canadian oil companies added to the rail network, which, by 2015, was capable of carrying just as much oil west or south to new terminals in the United States, China, and other world ports. Native resistance also keyed on the growing rail network, with mobilizations among the Quinault (on Washington State's Pacific Coast near one proposed terminal) and others. The amount of crude oil shipped by rail in the United States increased from carloads of 9,500 barrels in 2006 to 66,000 in 2011 and 234,000 in 2012 ("Lac-Mégantic" 2013).

"Over all," reported the *New York Times*, "Canada is poised to quadruple its rail-loading capacity over the next few years to as much as 900,000 barrels a day, up from 180,000 today . . . despite a derailment in the lakeside Quebec town of Lac-Mégantic in July [2013], in which a runaway oil train bound for a refinery in eastern Canada exploded, killing dozens of people and bankrupting the railway company" (Krauss 2013). "They don't give up," Jesse Prentice-Dunn, a Sierra Club policy analyst, said of the oil industry. Rail shipment is about $5 per barrel more expensive than pipelines. "Several Washington and Oregon refiners and ports are planning or building rail projects for Canadian heavy crude as well as light oil from North Dakota. The Texas refinery giant Tesoro and the oil services company Savage have announced a joint venture to build a $100 million, 42-acre oil-handling plant in the Port of Vancouver on the Columbia River that could handle 380,000 barrels of oil each day if permits are granted" (Krauss 2013).

### *THE QUINAULT RESIST OIL TRANSPORT*

After an oil train explosion killed 47 people on July 6, 2013, in Lac-Mégantic, Quebec, destroying the center of the town of 6,000 people, the Quinaults on the Pacific Coast of Washington State restated their opposition to the transport of oil and coal in their verdantly forested homeland. "It could have easily been Hoquiam" (a town south of the Quinault Nation), said Fawn Sharp, Quinault council president. "It is not a matter of 'if' these shipments will cause a major spill; it's a matter of 'when,' " said Sharp ("Lac-Mégantic" 2013). The Quinault oppose plans by Westway Terminal Company from Louisiana and Texas to construct a terminal for the shipment and storage of 800,000 barrels of oil in Grays Harbor near Hoquiam. The company's plans call for shipping and storing 10 million barrels of crude oil in that area.

"The massive train, oil barge and ship traffic this project will bring to Grays Harbor is a tragedy waiting to happen," Sharp said. "There will be spills and they will harm salmon, shellfish, and aquatic life, trample our treaty rights and cultural historic sites, and tie up traffic for extensive distances." Sharp added that the facility would violate Quinault treaty rights and their standards of "good stewardship and common sense. . . . The risk is not worth a few more, unsustainable jobs. Far too much is at stake, and there is simply no way oil train proponents can pass the straight face test and tell us that their proposal is safe. Lives are at stake. Fish and wildlife resources. Water quality and much, much more. These are the same type of rail cars that will come pouring through our area, and unquestionably threaten the lives and safety of our people and resources" ("Lac-Mégantic" 2013).

### *RESISTANCE TO FRACKING*

Fracking has sparked contention across Indian Country, from eastern New York State and Pennsylvania to New Brunswick, Canada, where 12 fracking protesters were arrested on June 21, 2013, on National Aboriginal Day for blocking work crews. Royal Canadian Mounted Police (RCMP) Corporal Chantal Farrah said, "They were attempting to block the heavy equipment from traveling on the road. . . . Now the people . . . were informed that they were breaking the law and that they needed to move. They refused and 12 people were arrested . . . seven men and five women" (Ball 2013b).

Native Americans are not opposing fracking alone, of course. By November 2013, more than 100 cities and towns in the United States had passed laws against fracking, according to a nonprofit monitor, FracTracker. Susan L. Brantley and Anna Meyendorff (2013) reported in the *New York Times* that "opposition to fracking has been considerable, if not unanimous, in the global green community, and in Europe in particular. France and Bulgaria, countries with the largest shale-gas reserves in Europe, have already banned fracking." Protesters blockaded some fracking test sites in the United Kingdom and Poland. In Britain, near Balcombe in West Sussex, up to 1,000 demonstrators set up a tent camp during the summer of 2013 to protest test drilling by the energy company Cuadrilla in what became a symbol of opposition to fracking. More than 100 people were arrested, including a member of Parliament from the Green Party, Caroline Lucas. "The company removed its test rig and left the site in late September," wrote Steven Erlanger (2013).

In 2011, Blood tribe women in Alberta set up a blockade against fracking on their lands to publicize their opposition. "As a grandmother and a mother, and as a member of the Blood Tribe, I have a vested interest like everyone that shares Mother Earth with us to help protect what we have," explained Cassie Brewer, holding up a sign decrying the practice. "The treaties pushed us here. We don't have another land to go to" (Ball 2013b).

Another fracking blockade was set up on the Elsipogtog First Nation of New Brunswick. On October 1, 2013, members and leaders of the Elsipogtog First Nation observed Mi'kmaq Treaty Day in Rexton on Provincial Route 134. Usually a day for emphasis on Mi'kmaq history and culture stemming from their 1752 pact with the British Crown, Treaty Day became an emphatic protest against fracking. Several dozen protesters blocked a road used by vehicles from SWN Resources Canada, a subsidiary of the Southwestern Energy Company, based in Houston, Texas. During the spring of 2012, SWN began seismic testing near the Elsipogtog First Nation. "Right now we're standing our ground and asking [SWN] to pull their equipment out of New Brunswick so that it will resolve peacefully," said John Levi, the warrior chief of Elsipogtog First Nation. "We're not going to back down" (Troian 2013a).

The blockade's members issued an "eviction notice" to frackers from Texas. Chief Arren Sock of the Elsipogtog First Nation served the notice, a Band Council resolution: "Whereas Prime Minister Harper and the Canadian Government have washed their hands with regards to the environmental protection of our lands and waters," read Chief Sock from a prepared statement, which also described lands that "have been assaulted by clear-cutting" as well as fracking, "we have lost all confidence in governments for the safekeeping of our lands held in trust by the British Crown." Their notice of eviction was "totally ignored by the provincial government and Southwestern Energy," said Sock. Thus, First Nations peoples "have been compelled to act and save our water, land and animals from ruin," so the "council of Elsipogtog are reclaiming all unoccupied reserved native lands back and put in the trust of our people. . . . I want a moratorium on fracking in New Brunswick . . . until there is such a time that they can come up with a safer solution on how to extract shale gas. Right now I am here for our children and their children's children. Right now nobody can guarantee a safe and effective way of extracting shale gas without harm to the environment. . . . No more negotiations with anybody" ("Elsipogtog Chief" 2013).

In the early morning of October 17, RCMP and local police clad in riot gear broke down the week-old barricade, firing tear gas and rubber bullets and arresting at least 40 people. The attack came after several days of relative peace. "For numerous days prior, RCMP were allowing first walking traffic, then one lane of automobile traffic, to pass freely through the blockaded area," wrote Miles Howe for the Halifax Media Co-op. "Anti-shale activists, as a measure of good faith, and in deference to emergency vehicles in particular, had days earlier removed two felled trees that had completely blocked off vehicular traffic" (Howe 2013).

Vincent Schilling (2013b), writing for the Indian Country Today Media Network, described the scene as follows:

> Chaotic: heavily armed Royal Canadian Mounted Police (RCMP) pouring into an encampment of sleeping protesters, leading dogs and carrying assault rifles. Amid burning police cars, pepper-spray-spewing hoses and barking police dogs, 28-year-old Amanda Polchies dropped to her knees, brandishing the only "weapon" she had an eagle feather. Holding it aloft, she began to pray An iPhone photo of Polchies was snapped, her feather aloft, in front of a wall of Royal Canadian Mounted Police, which went viral on social media, becoming an emblem of the protest.

"I saw elders getting doused with pepper spray. I also saw people's bruises from weapon-fired beanbags and rubber bullets, and police dogs and trucks all over the place. The dogs were barking so loud and jumping that these vehicles were shaking," said Michelin, an Inuit of mixed heritage from Labrador (Schilling 2013a). Some protesters, one of whom was pregnant (as well as a number of elders), faced off with private security guards and members of the RCMP who removed their blockade. Lorraine Clair, 44, according to one eyewitness, "was struck by an RCMP vehicle during an incident this week. She was also injured in an earlier confrontation with the RCMP," according to the Indian Country Today Media Network. "She's a fish plant worker, so her wrists were already weakened, and she was still healing from that, and then when they arrested her they broke her wrist," said protester Willie Nolan (Troian 2013b).

Howe, who was among those arrested, reported that "the ditch opposite me was already filled with 20-odd police in tactical blue uniforms, pistols already drawn. Three police officers dressed in full camouflage, one with a short-chained German Shepherd, were also near the ditch" (Howe 2013). "In the pre-dawn dark," wrote Howe, "about seven Molotov cocktails flew out of the woods opposite the police line stationed in the ditch. I cannot verify who threw these cocktails. They were—if it matters—lobbed ineffectively at the line of police and merely splashed small lines of fire across the road. A lawn chair caught fire from one cocktail. Two camouflaged officers then pumped three rounds of rubber bullet shotgun blasts into the woods." About 10 minutes after that, the confrontation turned into a melee after two officers had served the injunction on Seven Bernard, who was acting as a spokesman for the Warriors at the blockade, standing near a wall of officers with at least 60 guns drawn. As Sergeant Rick Bernard (no relation to Seven Bernard) began to serve the paper, Seven Bernard walked away. "Sergeant Bernard threw the injunction at his namesake, saying. 'Consider yourself served' " (Howe 2013).

Mi'kmaq Suzanne Patles attempted to intervene with a peace offering of tobacco as officers began hauling people away in handcuffs under a barrage of rocks and bottles. "Skirmishes then broke out in every direction," wrote Howe (2013). "From the highway side, District War Chief Jason Augustine was being chased by numerous police. In front of me, everywhere really, Warriors were being taken down by numerous RCMP officers in various clothes. Rubber bullet shots were fired by the RCMP, and both Jim Pictou and Aaron Francis both claim that they were hit—in the back and leg respectively" (Howe 2013). The Indian Country Today Media Network reported that "six RCMP vehicles were destroyed by fire, and several improvised explosive devices were discovered and defused," the RCMP said. "These explosives contained shrapnel and had the potential to seriously injure or kill people" ("Molotov" 2013). Some of the blockade members threw rocks and bottles at the police and the RCMP.

The Indian Country Today Media Network reported that "amateur photos and video have appeared online showing heavily armed police on the site and what appear to be snipers in nearby fields and forests. There are also photos of several police vehicles on fire. Those arrested, which may include elders conducting ceremonies at the site, are being detained by the Royal Canadian Mounted Police (RCMP). There were also some reports of shots fired" (Troian 2013b).

The RCMP said that the intervention and arrests were justified because the blockade had become a threat to public safety, while Canadian Assembly of First Nations National Chief Shawn Atleo said in a letter to New Brunswick Premier David Alward that a peaceful protest had been "disrupted through police and military intervention" (Stone 2013). News of the event spread rapidly on social media, and supporters began to arrive, some from as far away as Alberta. Floyd Augustine, a self-described leader of the Mi'kmaq Warrior Society, issued a call for support. Demonstrations were reported in Montreal, Vancouver, Ottawa, Iqaluit, Nunavut, Minneapolis, New York City, and San Francisco over the next few days as the Sierra Club announced support.

In an interview on Native Trailblazers Radio, Michelin said, "One of the hardest things for me to see was the treatment of Amy Sock [a member of the Elsipogtog First Nation]. She has a background in law and lives in Elsipogtog. This woman is one of the most peaceful people I have ever met in my life. There was a line of police standing around cars, and she came up with a white flag. She said she needed to see her people and wanted to make sure they were okay. She tried to break through the police line and started running. I saw people running behind her. She is about 125 pounds. I saw three big police officers tackle her to the ground like football players. They had her hands zip-tied, [pushing] her to the ground with her knee in her back, and the police officer looked up the same way a hunter poses over his kill. It broke my heart" (Schilling 2013a).

"We're not going to stop," said Sock. "We don't want shale gas to come to New Brunswick. . . . We have a big nuclear plant here. Once fracking goes on—once we start getting earthquakes—I'm afraid that thing is going to blow up. That's one of my fears. Our priority is Mother Earth." Photos of Sock spread rapidly when she was brutally arrested during the blockade. "I can cry, get mad, pray, forgive, and do my pipe ceremony to keep my strength and courage up," Sock said, "but this is one tough battle. We live by the river. . . . Our regular diet—fish, clams, eels, bass, salmon, and lobster—that's what we eat. I want to continue eating that without getting sick [from water pollution associated with fracking]." Reports by the Indian Country Today Media Network indicated that "despite some tribal leaders' support for shale gas exploration by SWN Resources Canada, which is preparing for seismic testing in Kent County, most Mi'kmaq are against the project" (Ball 2013b).

Until recent years, earthquakes were very rare in Oklahoma, which is not a seismically active area for natural reasons. Earthquakes, some as strong as 5.6 on the Richter scale, became more common, increasing an average of two a year (3.0 or greater) until 2008, to 20 in 2009, to 42 in 2010, and to 585 in 2014, triple the number in California. If all earthquakes are included, Oklahoma experienced more than 5,000 that year. Geologist William Ellsworth of the U.S. Geological Survey said at that time, "We can say with virtual certainty that the increased seismicity in Oklahoma has to do with recent changes in the way that oil and gas are being produced" (Galchen 2015, 35).

Nearly without exception, the stronger earthquakes are caused by fracking disposal wells, in which large amounts of brackish water (totaling billions of barrels) are pumped into the ground after having been brought up during underground oil shales mining. Other states (Ohio, Arkansas, Texas, Colorado, and Texas, among others) not noted for earthquakes have also been recording more quakes in fracked areas where wastewater has been poured back into the earth, although not in the numbers that by 2015 were shaking Oklahoma with an average of two earthquakes per day.

The fear of earthquakes is real. Research published in *the journal Science* "links four of Oklahoma's most prolific wastewater wells to a swarm of . . . small earthquakes near Jones, Oklahoma, by showing how the wells sent a wave of water pressure coursing through the subsurface. The pressure can reduce forces acting to keep faults locked and trigger earthquakes" (Hand 2014, 13). By 2014, Oklahoma, with its large concentration of fracking wells, was experiencing more earthquakes than California, "and seismologists are increasingly blaming them on the injection of wastewater from oil and gas operations" (Hand 2014, 13). Although "subsurface pressure data required to unequivocally link earthquakes to wastewater injection are rarely accessible," investigators writing in *Science* in 2014 linked Oklahoma earthquakes to fracking activity there (Keranen et al., 2014, 444–48).

"The Sierra Club stands with anti-fracking protesters in New Brunswick, Canada, and around the world who are protecting their land and their families from the real danger that fracking brings to the health and safety of their communities," said Sierra Club Executive Director Michael Brune. "All Canadians and all Americans should ask themselves whether a police response with tactical units and snipers was meant to serve public safety, or squelch opposition to fracking in the service of the oil industry" ("Molotov" 2013). "We are shocked by yesterday's developments and we pray for the safety of Chief Aaren Sock, his community members and other land defenders who are at the site on Elsipogtog First Nation traditional lands," said Stan Beardy, head of the Chiefs of Ontario as well as the regional chief for the Assembly of First Nations. "It is past time now to call a halt to the physical exploration work and engage Elsipogtog First Nation in a respectful dialogue. In my view, this course is in the best interests of everyone and all concerned" ("Molotov" 2013).

## *FRACKING LINKED TO WATER POLLUTION*

Protesters mounted their blockades following reports of water pollution associated with fracking. For example, in 2012, contaminants linked to fracking were detected by the U.S. Geological Survey in the groundwater of Pavillion, Wyoming, on the Wind River Indian reservation. The contamination was detected in one of two wells drilled by the EPA in 2010. Residents of Pavillion had reported that the water had a foul smell and taste. The Indian Country Today Media Network reported that "the USGS announced that the water had again tested positive for high levels of methane, ethane, diesel compounds and phenol . . . some of the chemicals used for fracking" ("Wyoming Groundwater" 2012).

Encana Corporation, a drilling company based in Calgary, asserted that the EPA had drilled into a gas reserve. The company seemed to have no explanation for the ill-smelling and ill-tasting tap water. As of 2013, Encana had 140 natural gas wells in the area. "This goes to the heart of concerns raised by state and federal agencies, as well as Encana: EPA's wells are improperly constructed," Encana spokesman Doug Hock told the *Wall Street Journal* ("EPA Says" 2012).

The *New York Times* reported that with unemployment reaching 70 percent during some seasons, a debate has broken out among the Blackfeet "whether the sacredness of the mountains, streams and vegetation can be maintained if the land is drilled and its oil and gas resources extracted" ("The New York Times" 2012). With at least 30 test wells being drilled, employing as many as 45 to 50 people each, the money is tempting. Pauline Matt told the *Times* that the drilling "threatens everything we are as Blackfeet" (Healy 2012).

"The Blackfeet are one of many tribes facing such choices," the Indian Country Today Media Network reported ("The New York Times" 2012). The Turtle

Mountain Band of Chippewa Tribal Council banned fracking on its reservation following a presentation to its Tribal Council by the No Fracking Way Turtle Mountain tribe. Carol Davis, a member of the group, called its presentation "Fracking 101." "I know there's an oil boom and it's providing a lot of jobs," she added, "but we can't risk contaminating our water on the Turtle Mountain Reservation for the sake of money" ("Turtle Mountain" 2012). The Turtle Mountain band's lands, in Rolette County, North Dakota, lie atop part of the Bakken shale formation.

## *OIL AT FORT BERTHOLD*

"I wanted to write a story about strength and resilience. I wanted to write a story about the singers, the horse people, and the earth lodge builders of the Mandan, Hidatsa, and Arikara peoples, the squash and corn, the heartland of agricultural wealth in the Northern Plains," wrote Anishinaabe (Ojibway) environmental activist and writer Winona LaDuke. Instead, on the northern Missouri River, she found a land remade by the Bakken oil boom. By 2014, in six years, U.S. oil production increased 70 percent as imports from the Organization of Petroleum Exporting Countries fell by half (Krauss 2014, B-3). Today, what remains of the Mandan, Hidatsa, and Arikara's land base makes up the Fort Berthold reservation, a grid-work of wells—about 1,370 wells on the reservation, drilled and fracked, as of late 2014 (Sontag and McDonald 2014). Where "the 1954 Garrison Diversion project . . . submerged a people under Lake Sakakawea, taking 152,000 acres of their best land . . . drown[ing] their villages, . . . agricultural wealth [and] history," the tribal council, now materially wealthy from oil and casino revenues, hosts oil executives and politicians on its 149-passenger yacht (LaDuke 2014b). The 96-foot yacht, named *Island Girl*, which cost about $2.5 million, spends most of its time out of the water, sitting on blocks, where it has become "a symbol to many of their leaders' misplaced priorities" (Sontag and McDonald 2014).

So far, at least 1,370 wells have been drilled and fracked, ar Fort Berthold. By the end of 2014, the wells were pumping more than 386,000 barrels of oil a day, one-third of North Dakota's Bakken output. According to the *New York Times*, there were 850 oil-related environmental incidents at Fort Berthold reported by companies from 2007 through mid-October 2014. "Our tribal council is so focused on money, money, money," said Edmund Baker, the reservation's environmental director. "And our tribal chairman [Tex G. Hall] is: 'Edmund, don't tell me about spills. I'm busy trying to do things for my people' " (Sontag and McDonald 2014).

Tribal activists on the Fort Berthold reservation, such as Kandi Mossett, now describe a landscape that would be unrecognizable to their ancestors, including not only a maze of oil derricks but also "the huge Basin Electric coal generation facilities, burning the dirtiest coal in America, just upwind from their villages, [and] oil refinery proposals that have been accelerated through federal processes," along with fracking, with its appetite for precious water, as it pollutes and irradiates more than 30 trillion gallons of toxic liquid deep underground. "Simply stated," wrote LaDuke (2014b), "once water has been used in fracking, it is no longer living water. It is dead, and it is lethal." Some of this "dead" water spills onto the surface. On July 8, 2014, in what became known as the Crestwood spill, about a million gallons of saline, radioactive water leaked out of a pipe and, moving toward Lake Sakakawea, was blocked only by three beaver dams.

Despite the new wealth, many people's quality of life has deteriorated. "Apart from a significant rise in jobs, which often go to transient workers," Deborah Sontag and Brent McDonald (2014) wrote in the *New York Times*,

> many see deterioration rather than improvement in their standard of living. They endure intense truck traffic, degraded roads, increased crime, strained services and the pollution from spills, flares and illegal dumping. . . . Deep-seated problems can be hard to fix—a life expectancy of 57, for instance, compared with 79 for North Dakotans as a whole. But its critics say the tribal government has invested little in social welfare, like desperately needed housing, and has distributed little of the $200 million set aside in the People's Fund.

"Every single day more than 100 million cubic feet of natural gas is flared away," Fort Berthold tribal member and antifracking activist Mossett said. "That's enough to heat half a million homes. That's as much carbon dioxide emitted as 300,000 cars. That's crazy." The gas, which could be used as fuel, is being flared because the area lacks the required infrastructure to move it. As much as 70 percent of gas from wells on the reservation is flared (LaDuke 2014a).

## *PROTESTS OF PIPELINES ACROSS CANADA*

Paralleling the debate over the Keystone XL Pipeline in the United States, environmentalists and representatives from more than 160 First Nations across Canada united in opposition to proposed pipelines (or expansions of existing ones) that would carry increasing production from Alberta's tar sands fields westward and eastward for export. The issues are the same as in the United States: locally, the danger of spills, and worldwide, the atmospheric pollution posed by a new source of carbon dioxide.

The Northern Gateway Pipeline, proposed by Enbridge, was being designed to carry a daily average of roughly 525,000 barrels of diluted bitumen from the oil sands fields of Bruderheim, Alberta, to a planned marine terminal at Kitimat, British Columbia, near Vancouver, bound for export to Asia by ship. Environmentalists assert that existing Enbridge pipelines have been responsible for 801 spills in 10 years ("Defend" 2013). The U.S. oil company Kinder Morgan also was planning to spend $5 billion, tripling capacity of the Trans Mountain Pipeline from the tar sands fields to the Vancouver area.

TransCanada also has proposed an Energy East pipeline to carry about 1.1 million barrels of crude a day from Saskatchewan and Alberta and through the Quebec City and Montreal areas to the vicinity of St. John, New Brunswick. Another pipeline, which now carries natural gas, would be converted to export diluted bitumen. TransCanada announced in 2013 that it planned to spend $12 billion on these projects.

In 2013, according to the Environment News Service, "polls show[ed] 80 percent of British Columbians opposed to crude oil tanker traffic on the BC coast," as there was an "unbroken wall of opposition" of indigenous nations, "with more than 160 First Nations putting their signatures on the Save the Fraser Declaration opposing tar sands exports projects through their territories" ("Defend" 2013). On November 16 of that year, more than 2,000 people demonstrated in Vancouver against the Northern Gateway project in a "Defend Our Climate" rally.

The proposed expansion of Kinder Morgan's tar sands oil pipeline was strenuously criticized by Native peoples in Washington State and British Columbia on October 22, 2014, during hearings before Canada's National Energy Board. A pipeline terminal also is being considered at Cherry Point in far northwestern Washington State, a few miles south of Vancouver. Expanding the pipeline would increase tanker loadings and put tribal fishing people at risk, "not to mention drastically increase the chance of a catastrophic oil spill," Glen Gobin, a member of the Tulalip Board of Directors, told the National Energy Board (Hansen 2014).

A caravan of Canadian First Nations people opposed to Enbridge's Northern Gateway Pipeline, led by the Yinka Dene Alliance, formed a "Freedom Train" from Jasper, Alberta, to Toronto, that departed on April 30, 2012, to rally at Enbridge's annual meeting on May 9. The 30 native nations supporting the caravan of about 40 people said that the trip was "aimed at protecting their rights and their freedom from the threat of devastating oil spills" ("Freedom Train" 2012). "Our journey on this Freedom Train is to build on the support from the people of Canada for the protection of our traditional lands. We will stand firm against all industries like Enbridge that are planning to destroy the environment and the future of all peoples," said Chief Martin Louie of the Nadleh Whut'en First Nation, a member of the Yinka Dene Alliance, in the statement. "We need a healthy environment to ensure a healthy future for our children" ("Freedom Train" 2012).

"We're traveling across Canada to tell Enbridge that they will not be permitted to build their pipelines through our lands, period," said Chief Jackie Thomas of Saik'uz First Nation. "The fight against Enbridge is a fight for our freedom to govern ourselves and to choose our own future. We will not accept the government imposing a decision on us and forcing this pipeline through our lands" ("Freedom Train" 2012).

The government of British Columbia rejected the Gateway Pipeline, having "thoroughly reviewed all of the evidence and submissions made to the panel and asked substantive questions about the project including its route, spill response capacity and financial structure to handle any incidents," said provincial Environment Minister Terry Lake in a statement. "Our questions were not satisfactorily answered during these hearings" ("Freedom Train" 2012). "The B.C.

government's detailed submittal to federal government decision-makers was also noteworthy in recognizing there is growing evidence that diluted bitumen could pose additional risks to water and is more difficult to clean up," the Natural Resources Defense Council said in a statement. "In light of British Columbia's rejection, the U.S. State Department should revisit its previous findings that the Keystone XL poses no risks to water or climate which were based on faulty assumptions about the behavior of tar sands oil and about the role that Keystone XL will play as the major driver of tar sands strip-mining and drilling" ("Freedom Train" 2012).

In early January 2012, Canadian Prime Minister Stephen Harper accused "foreign-funded" radicals of stoking opposition to the Northern Gateway Pipeline, even as much of the opposition came from indigenous people inside the country. Harper said that Tides Canada, an environmental group, had received donations from the United States. "These groups threaten to hijack our regulatory system to achieve their radical ideological agenda," wrote Minister of Natural Resources Joe Oliver in an open letter a few days later. "They use funding from foreign special interest groups to undermine Canada's national economic interest" ("Ottawa Blasts" 2012).

"The minister's allegations about radicals using foreign money to achieve an ideological agenda were sweeping, and we assume he was referring to the Dene Nation and other First Nations and aboriginal organizations participating in the review process," said Dene National Chief Bill Erasmus in a statement responding to Oliver's letter. "Our mandate is to preserve and protect our communities, our land, and our culture, and it is our democratic right to participate in hearings on a pipeline that will impact us" ("Ottawa Blasts" 2012). "We have great concern about Northern Gateway and tankers coming through our traditional waters," Frank Brown, of Heiltsuk First Nation in Bella Bella, told the Indian Country Today Media Network. "If necessary, we will take both legal means and other means necessary to ensure we protect the traditional territories from which we come from the beginning of time" (Ball 2013a).

At the initial hearing on January 14, 2013, in Vancouver, more than 1,000 protesters gathered outside. Inside, several speakers from Canadian First Nations and environmental groups asserted that tanker or pipeline accidents could threaten the coastline as well as nearby rivers and lands. When people made noise, the government closed the hearing to the public. Activists who snuck past police were arrested for assault or trespass, according to the Indian Country Today Media Network. "The indigenous people across whose sovereign, unceded territories this pipeline is supposed to cross have already said no, unequivocally," said Kim Heartty, who was arrested. "So the project should not be under consideration beyond that" (Ball 2013a).

"Protecting people and the environment is our top priority, which is why we announced enhancements to make what was already a very safe project even safer," said Janet Holder, an Enbridge executive vice president. "We intend to demonstrate to British Columbians and to all Canadians through the examination of the facts and science upon which this project application is based that there is a path forward that provides for prosperity while protecting the environment" (Ball 2013a).

In December 2013, Canada's federal National Energy Board and the Canadian Environmental Assessment Agency approved the Northern Gateway Pipeline, setting it up for expected approval by Canada's conservative-dominated cabinet, probably to be followed by a confrontation with British Columbia's liberal provincial government as well as First Nations (whose lands may be traversed by the pipeline) and environmental groups. The review panel (which has three members) said that a major oil spill that would "have significant adverse environmental effects on ecosystems" is not likely and that "any harm . . . could be cleaned up and affected areas returned to a functioning ecosystem similar to that existing prior to the spill. . . . After weighing all the oral and written evidence, the panel found that Canada and Canadians would be better off with the Enbridge Northern Gateway project than without it" (Austen 2013).

In June 2014, the Canadian federal government's approval of the Enbridge Northern Gateway Pipeline united several hundred First Nations leaders and environmental groups in a pledge to take up legal action to stall the project. They indicated that international human rights law would be used as a litigating agent.

"The First Nations Leadership Council, which is composed of the B.C. Assembly of First Nations, First Nations Summit and Union of B.C. Indian Chiefs, is completely disgusted at this decision," said the organizations in a joint statement on June 17, 2014. "There is an undeniable and inherent risk attached to this project and the idea of a catastrophic ecological disaster is unacceptable for the people of this Province. Delaying this project will only serve to fortify the opposition to this project," said Grand Chief Stewart Phillip, president of the Union of B.C. Indian Chiefs ("Canada Approves" 2014).

A coalition said that "international human-rights standards, as set out in the *United Nations Declaration on the Rights of Indigenous Peoples* and the rulings of regional and international human rights bodies, also require a high standard of precaution in all decisions affecting Indigenous peoples' rights and their lands, territories and resources. In many instances, the standard required is that projects should proceed only with the free, prior and informed consent of Indigenous peoples" ("Canada Enbridge" 2014).

Yet another pipeline, Pacific Trails, became an object of resistance on November 27, 2012, by indigenous peoples from California and several major Canadian cities as well as in Port of Spain, Trinidad and Tobago (outside the Canadian embassy) a few days after hereditary leaders of the Wet'suwet'en Nation evicted surveyors working on the pipeline's proposed route. This pipeline was being planned to carry 1 million cubic feet of natural gas per day obtained through fracking 290 miles from Summit Lake, British Columbia, to Kitimat on the Pacific Coast, the same terminus as the Enbridge Northern Gateway Pipeline. Company representatives told the media that they had indigenous support along the pipeline route, but hereditary, traditional leadership had been largely ignored.

The Wet'suwet'en Nation leaders issued a traditional warning to remove trespassers from a nation's traditional territory. Contractors' equipment also was confiscated. "If the Pacific Trails Pipeline decides to push their agenda, along with the federal government and provincial government, to try to force this pipeline through our lands, they're going to continue to meet us, and we're going to keep resisting them," Toghestiy, a hereditary chief of Wet'suwet'en Nation, told the Indian Country Today Media Network. "If they decide to escalate it, we'll have to do the same. It's something that we don't want to do, but if they're not willing to sit down and have meaningful consultation with our hereditary chiefs, and with the Wet'suwet'en people, then they're going to be meeting a lot of resistance up here" (Ball 2012c).

Opposition to Kinder Morgan's proposed pipeline by late 2014 focused on Burnaby Mountain, a conservation area. Beginning with a camp of a few people in September, the protest grew to more than 1,000 by the end of November, most of whom ignored police orders to disperse. More than 120 were arrested on civil disobedience charges when they crossed police lines. First Nations peoples joined non-Indians in the protest. On November 27, a British Columbia Supreme Court judge "refused the company's request to extend [an] injunction against disruptive protests through December 12, effectively ordering Kinder Morgan to clear its equipment out of the Burnaby Mountain Conservation Area" (Ball 2014b).

People from the Heiltsuk First Nation in Bella Bella, British Columbia, arrived, declaring "unwavering support for the keepers of the sacred fire on Burnaby Mountain. . . . Our relatives from the Salish Sea are demonstrating the courage and conviction of our ancestors," said Heiltsuk Chief Councilor Marilyn Slett. "Our peoples have always been keepers of the lands and waters," added Heiltsuk Hereditary Chief Harvey Humchitt. "Now we are called to be defenders as well as keepers. The land defenders on Burnaby Mountain are upholding ancestral law first and foremost, and we support them wholeheartedly" ("Tar Sands Opposition" 2014). City officials in Burnaby, a suburb of Vancouver, have sued Kinder Morgan even as the company began to test the mountain for construction in three places, drilling holes and felling trees.

The Union of British Columbia Indian Chiefs (UBCIC) also expressed opposition to the pipeline, saying that it "stands in solidarity with those that have been arrested." Union President and Grand Chief Stewart Phillip said that "we will continue to stand in support with those on the Mountain to uphold and defend Indigenous rights, land rights and human rights. . . . It is infuriating and beyond frustrating that we are faced with this provocative and heavy-handed approach by the RCMP when at this time the City of Burnaby's court proceedings have not even been completed. Kinder Morgan is despoiling the Burnaby Mountain Conservation Area to brazenly push ahead with their proposed expanded pipeline in the face of massive opposition" ("Tar Sands Opposition" 2014).

## *FURTHER READING*

"Anishinaabe and Lakota Riders Protest Pipelines, on Horseback." 2013. Indian Country Today Media Network, October 18. http://indiancountrytodaymedianetwork

.com/gallery/photo/anishinaabe-and-lakota-riders-protest-pipelines-horseback-151792

"Athabasca Oil Sands." 2011. NASA Earth Observatory, November 30. http://earthobservatory.nasa.gov/IOTD/view.php?id=76559&src=eoa-iotd

Austen, Ian. 2013. "Canadian Review Panel Approves Plans for an Oil Pipeline." *New York Times*, December 19. http://www.nytimes.com/2013/12/20/business/international/canadian-review-panel-approves-plans-for-an-oil-pipeline.html

Ball, David P. 2012a. "Athabasca Chipewyan Launch Treaty 8 Challenge to Shell Canada over Oil Sands." Indian Country Today Media Network, October 5. http://indiancountrytodaymedianetwork.com/article/athabasca-chipewyan-launch-treaty-8-challenge-to-shell-canada-over-oil-sands-137632

Ball, David P. 2012b. "Spiritual Leaders Vow to Defend Mother Earth from Oil Sands and Pipelines with Spiritual Declaration." Indian Country Today Media Network, November 23. http://indiancountrytodaymedianetwork.com/spiritual-leaders-vow-to-defend-mother-earth-from-oil-sands-and-pipelines-with-spiritual-declaration-147332

Ball, David P. 2012c. "U.S. and Canada-Wide Protests Target Pacific Trails' Proposed Fracking Pipeline." Indian Country Today Media Network, November 27. http://indiancountrytodaymedianetwork.com/article/us-and-canada-wide-protests-target-pacific-trails-proposed-fracking-pipeline-145895

Ball, David P. 2013a. "Enbridge Faces Rising Opposition to Northern Gateway Pipeline with Protests, Arrests." Indian Country Today Media Network, January 25. http://indiancountrytodaymedianetwork.com/2013/04/04/red-lake-band-chippewa-determined-blockade-atop-four-enbridge-pipelines-minnesota-148600

Ball, David P. 2013b. Fracking Troubles Atlantic First Nations after Two Dozen Protesters Arrested. Indian Country Today Media Network, June 28. http://indiancountrytodaymedianetwork.com/2013/06/28/fracking-troubles-atlantic-first-nations-after-two-dozen-protesters-arrested-150192

Ball, David. 2014a. "Neil Young: Blood of First Nations People Is on Canada's Hands." Indian Country Today Media Network, January 14. http://indiancountrytodaymedianetwork.com/2014/01/14/neil-young-blood-first-nations-people-canadas-hands-153104

Ball, David P. 2014b. "Tensions Simmer after Kinder Morgan Confrontation over Pipeline Extension in Burnaby, B.C." Indian Country Today Media Network, December 12, 2014. http://indiancountrytodaymedianetwork.com/2014/12/12/tensions-simmer-after-kinder-morgan-confrontation-over-pipeline-extension-burnaby-bc

Berry, Carol. 2012a. "Alberta Oil Sands Up Close: Gunshot Sounds, Dead Birds, a Moonscape." Indian Country Today Media Network, February 2. http://indiancountrytodaymedianetwork.com/article/alberta-oil-sands-up-close%3A-gunshot-sounds%2C-dead-birds%2C-a-moonscape-95444

Berry, Carol. 2012b. "Oglalas' Keystone XL Pipeline Position Issued." Indian Country Today Media Network, January 31. http://indiancountrytodaymedianetwork.com/article/oglalas%E2%80%99-keystone-xl-pipeline-position-issued-95128

Berry, Carol. 2013. "Tribal Members Sign Treaty Calling for an End to Alberta Oil Sands Development and Keystone XL." Indian Country Today Media Network, January 31. http://indiancountrytodaymedianetwork.com/2013/01/31/tribal-members-sign-treaty-calling-end-alberta-oil-sands-development-and-keystone-xl

"Blood Tribe Members Set Up Anti-Fracking Blockade." 2011. Indian Country Today Media Network, September 10. http://indiancountrytodaymedianetwork.com/article/blood-tribe-members-set-up-anti-fracking-blockade-53321

Brantley, Susan L., and Anna Meyendorff. 2013. "The Facts on Fracking." *New York Times*, March 13. http://www.nytimes.com/2013/03/14/opinion/global/the-facts-on-fracking.html

"Call to Action! Rally at Cuomo's State of the State: No Fracking, Yes Renewables!" 2014. Two Row Wampum Renewal Campaign, January 14. http://honorthetworow.org/call-to-action-rally-at-cuomos-state-of-the-state-no-fracking-yes-renewables

"Canada Approves Enbridge Pipeline through B.C., First Nations Will Sue." 2014. Indian Country Today Media Network, June 19. http://indiancountrytodaymedianetwork.com/2014/06/19/canada-approves-enbridge-pipeline-through-bc-first-nations-will-sue-155377

"Canada Enbridge Approval Unites First Nations, Amnesty, Rights Groups in Opposition." 2-14. Indian Country Today Media Network, June 20. http://indiancountrytodaymedianetwork.com/2014/06/20/canada-enbridge-approval-unites-first-nations-amnesty-rights-groups-opposition-155400

Capriccioso, Rob. "Indigenous Oil Sands Protest Leads to White House Arrests." Indian Country Today Media Network. September 2, 2011. http://indiancountrytodaymedianetwork.com/article/indigenous-oil-sands-protest-leads-to-white-house-arrests-51420

"Chiefs Declare Keystone XL Consultation Meeting Invalid, Walk Out on State Department Officials." 2013. Indian Country Today Media Network, May 17. http://indiancountrytodaymedianetwork.com/2013/05/17/chiefs-declare-keystone-xl-consultation-meeting-invalid-walk-out-state-department

Dadigan, Marc. 2013. "Umatilla Tribe Battles Mega-Loads Headed for Alberta Oil Sands." Indian Country Today Media Network, December 1. http://indiancountrytodaymedianetwork.com/2013/12/11/umatilla-tribe-battles-mega-loads-headed-alberta-oil-sands-152657

" 'Defend Our Climate' Rallies Draw Thousands across Canada." 2013. Environment News Service, November 18. http://ens-newswire.com/2013/11/18/defend-our-climate-rallies-draw-thousands-across-canada

"Elsipogtog Chief Issues Eviction Notice to Texas-Based Frackers. Band Council Resolution to Reclaim All Unoccupied Crown Land." 2013. Halifax Media Co-op, October 4. http://halifax.mediacoop.ca/story/elsipogtog-chief-issues-eviction-notice-texas-base/1909

"Emissions from Oil Sands Mining." 2012. NASA Earth Observatory, March 2. http://earthobservatory.nasa.gov/IOTD/view.php?id=77283&src=eoa-iotd

"EPA Says Wyoming Fracking Results Are Consistent." 2012. *Wall Street Journal*, September 26. http://online.wsj.com/article/SB10000872396390443328404578020923049282436.html

Erlanger, Steven. 2013. "As Drilling Practice Takes Off in U.S., Europe Proves Hesitant." *New York Times*, October 9. http://www.nytimes.com/2013/10/10/world/europe/as-drilling-practice-takes-off-in-us-europe-proves-hesitant.html

" 'Every Tree and Plant Died': Massive Toxic Spill Guts Alberta." 2013. Indian Country Today Media Network, August 26. http://indiancountrytodaymedianetwork.com/2013/08/26/toxic-wastewater-spill-alberta-kills-dene-tha-landscape-150968

"Fill Gaps in Keystone XL Draft Environment Report or Reject Pipeline, NCAI Tells Obama Administration." 2013. Indian Country Today Media Network, http://indiancountrytodaymedianetwork.com/2013/05/02/fill-gaps-keystone-xl-draft-environment-report-or-reject-pipeline-ncai-tells-obama.html

"'Freedom Train' Protesting Northern Gateway Pipeline Crossing Canada." 2012. Indian Country Today Media Network, April 30. http://indiancountrytodaymedia network.com/most-popular?page=2677

Frosch, Dan. 2013. "Amid Pipeline Debate, Two Costly Cleanups Forever Change Towns." *New York Times*, August 10. http://www.nytimes.com/2013/08/11/us/amid-pipeline-debate-two-costly-cleanups-forever-change-towns.html.

"Full Text of the Great Plains Tribal Chairmen's Association Statement against the Keystone XL Pipeline." 2013. Indian Country Today Media Network, May 17. http://indiancountrytodaymedianetwork.com/2013/05/17/full-text-great-plains-tribal-chairmens-association-statement-against-keystone-xl

Galchen, Rivka. 2015. "Weather Underground." *The New Yorker*, April 13, 34–40.

Gillis, Justin, and Henry Fountain. 2014. "Trying to Reclaim Leadership on Climate Change." *New York Times*, June 2, A-12.

Hand, Eric. 2014. "Injection Wells Blamed in Oklahoma Earthquakes." *Science* 345 (July 4): 13–14.

Hansen, James. 2012. "Game Over for the Climate." *New York Times*, May 9. http://www.nytimes.com/2012/05/10/opinion/game-over-for-the-climate.html

Hansen, Terri. 2014. "Coast Salish Unite against Tripling Capacity of Kinder Morgan Tar Sands Pipeline." Indian Country Today Media Network, October 27. Indian countrytodaymedianetwork.com/2014/10/27/coast-salish-unite-against-tripling-capacity-kinder-morgan-tar-sands-pipeline-157543

Healy, Jack. 2012. "Tapping into the Land, and Dividing Its People." *New York Times*, August 15. http://www.nytimes.com/2012/08/16/us/montana-tribe-divided-on-tapping-oil-rich-land.html

Homer-Dixon, Thomas. 2013. "The Tar Sands Disaster." *New York Times*, March 31. http://www.nytimes.com/2013/04/01/opinion/the-tar-sands-disaster.html

Howe, Miles. 2013. "RCMP Bring 60 Drawn Guns, Dogs, Assault Rifles, to Serve Injunction on the Wrong Road." Halifax Media Co-op, October 18. http://halifax.mediacoop.ca/story/rcmp-bring-60-drawn-guns-dogs-assault-rifles-serve/19358

"In Historic Visit, Obama Vows to Help." 2014. *Omaha World-Herald*, June 14, A-1, A-3.

"Iñupiat Hunting Community Divided over Shell Drilling in Chukchi Sea off Alaska." 2012. Indian Country Today Media Network, October 4. http://indiancountrytodaymedianetwork.com/article/i%C3%B1upiat-hunting-community-divided-over-shell-drilling-in-chukchi-sea-off-alaska-137125

Johnson, Kirk. 2013. "Fight over Energy Finds a New Front in a Corner of Idaho." *New York Times*, September 25. http://www.nytimes.com/2013/09/26/us/fight-over-energy-finds-a-new-front-in-a-corner-of-idaho.html

"Just Say No Mega-Load! Columbia River Inter-Tribal Fish Commission Joins Protest." 2013. Indian Country Today Media Network, September 9. http://indiancountrytodaymedianetwork.com/2013/09/09/columbia-river-inter-tribal-fish-commission-joins-anti-mega-load-chorus-resolution-151203

Keranen, K. M., M. Weingarten, G. A. Abers, B. A. Berkins, and S. Ge. 2014. "Sharp Increase in Central Oklahoma Seismicity since 2008 Induced by Massive Wastewater Injection." *Science* 345 (July 25): 444–48.

Klein, Naomi. 2014. *This Changes Everything: Capitalism and the Climate*. New York: Simon & Schuster.

Koch, Wendy. 2014. "Would Keystone Pipeline Unload 'Carbon Bomb' or Job Boom?" *USA Today*, March 10. http://www.usatoday.com/story/news/nation/2014/03/01/keystonexls-myths-debunked/5651099

Kolbert, Elizabeth. 2007. "Unconventional Crude: Canada's Synthetic-Fuels Boom." *The New Yorker*, November 12, 46–51.

Krauss, Clifford. 2013. "Looking for a Way around Keystone XL, Canadian Oil Hits the Rails." *New York Times*, October 31. http://www.nytimes.com/2013/10/31/business/energy-environment/looking-for-a-way-around-keystone-xl-canadian-oil-hits-the-rails.html

Krauss, Clifford. 2014. "U.S. Oil Prices Fall below $80 a Barrel." *New York Times*, November 4, B-3.

Krauss, Clifford, and Ian Austen. 2014. "Rocky Road for Canadian Oil." *New York Times*, May 13, B-1, B-4.

Krauss, Clifford, and Jad Mouawad. 2014. "Accidents Surge as Oil Industry Takes the Train." *New York Times*, January 26, A-1.

"Lac-Mégantic Rail Tragedy Resonates at Quinault Nation." 2013. Indian Country Today Media Network, July 29. http://indiancountrytodaymedianetwork.com/2013/07/29/lac-megantic-rail-tragedy-resonates-quinault-nation-victims-are-memorialized-150626

LaDuke, Winona. 2014a. "Bakken Gas Flares Away, as Nationwide Propane Shortage Kills with Hypothermia." Indian Country Today Media Network, October 9. http://indiancountrytodaymedianetwork.com/2014/10/09/bakken-gas-flares-away-nationwide-propane-shortage-kills-hypothermia-157249

LaDuke, Winona. 2014b. " 'Unspeakable Poverty of Loss': Intergenerational Trauma and the Bakken Oil Fields." Indian Country Today Media Network, October 8. http://indiancountrytodaymedianetwork.com/2014/10/08/unspeakable-poverty-loss-intergenerational-trauma-and-bakken-oil-fields-157243

"Leo DiCaprio Tours Tar Sands, Joins Natives in ALS Ice Bucket Challenge." 2014. Indian Country Today Media Network, August 26. http://indiancountrytodaymedianetwork.com/2014/08/26/leo-dicaprio-tours-tar-sands-joins-natives-als-ice-bucket-challenge-156602

Leslie, Jacques. 2014. "Is Canada Tarring Itself?" *New York Times*, March 31, A-19.

Lizza, Ryan. 2013. "The President and the Pipeline." *The New Yorker*, September 16, 38–51.

"Lower Brule Sioux Reject Keystone XL, Evict TransCanada." 2015. Indian Country Today Media Network, May 1. http://indiancountrytodaymedianetwork.com/2015/05/01/lower-brule-sioux-reject-keystone-xl-evict-transcanada-160214

Malakoff, David. 2014. "Oil Sands Fight Heats Up in U.S." *Science* 345 (July 25): 368–69.

McLinden, C. A., V. Fioletov, K. F. Boersma, N. Krotkov, C. E. Sioris, J. P. Veefkind, and K. Yang. 2012. "Air Quality over the Canadian Oil Sands: A First Assessment Using Satellite Observations." *Geophysical Research Letters* 39: L04804.

McNeel, Jack. 2015. "Too Many Broken Pipelines: Fort Peck Reservation Passes Resolution Opposing Keystone XL through Montana." Indian Country Today Media Network, February 24. Indiancountrytodaymedianetwork.com/2015/02/24/too-many-broken-pipelines-fort-peck-reservation-passes-resolution-opposing-keystone-xl

"Message in a Blanket: Dakota Artist Stealth-Embroiders 'NOKXL' into Obama Quilt." 2015. Indian Country Today Media Network, May 16. http://indiancountrytodaymedianetwork.com/2015/05/16/message-blanket-dakota-artist-embroiders-nokxl-obama-quilt-160385

"Molotov Cocktails and Guns Confiscated; Support Pours in for Mi'kmaq Protesters." 2013. Indian Country Today Media Network, October 18. http://indiancountry

todaymedianetwork.com/2013/10/18/support-pours-mikmaq-anti-fracking-protesters-new-brunswick-events-spring-and-sierra-club

Morin, Brandi. 2013. "Federal and Provincial Governments Lose Appeal against First Nation Oil Sands Lawsuit." Indian Country Today Media Network, May 8. http://indiancountrytodaymedianetwork.com/2013/05/08/federal-and-provincial-governments-lose-appeal-against-first-nation-oil-sands-lawsuit

"NASA Scientist, Religious Leaders Arrested in Tar Sands Protest." 2011. Environment News Service, August 29. http://www.ens-newswire.com/ens/aug2011/2011-08-29-02.html

"NCAI Condemns Keystone XL Pipeline." 2011. Indian Country Today Media Network, August 18. http://indiancountrytodaymedianetwork.com/article/ncai-condemns-keystone-xl-pipeline-48112

"The *New York Times* Profiles Blackfeet Drilling Dilemma." 2012. Indian Country Today Media Network, August 16. http://indiancountrytodaymedianetwork.com/article/%3Cem%3Ethe-new-york-times%3C/em%3E-profiles-blackfeet-drilling-dilemma-129595

"Nez Perce Victory: U.S. Forest Service Forbids Mega-Loads along Highway 12." 2013. Indian Country Today Media Network, September 20. http://indiancountrytodaymedianetwork.com/2013/09/20/nez-perce-victory-us-forest-service-forbids-mega-loads-along-highway-12-151372

Nikiforuk, Andrew. 2014. "Large Dams of Mining Waste Leaking into Athabasca River: Study." Indian Country Today Media Network, February 26. Indiancountrytodaymedianetwork.com/2014/02/26/large-dams-mining-waste-leaking-athabasca-river-study-153749?page=0%2C2

"Ottawa Blasts 'Radicals,' Blames 'Foreign Interests' for Northern Gateway Opposition." 2012. Indian Country Today Media Network, March 14. http://indiancountrytodaymedianetwork.com/article/ottawa-blasts-%27radicals%2C%27-blames-%27foreign-interests%27-for-northern-gateway-opposition

"Protecting Knowledge: Native Youth Movement Protest at Sun Peaks Resort; Interview with Amanda Soper, Spokesperson for the Native Youth Movement, and Chris Rogers, Spokesperson for Sun Peaks Resort." 2001. CBC Radio, B.C. Almanac, August 24.

Ross, John. 1996. "Is Zapatista Rebellion Rooted in Oil?" *Earth Island Journal* 11, no. 2: 20.

Schilling, Vincent. 2012. "Pine Ridge Residents Halt Canadian Mine Equipment Transportation through Reservation." Indian Country Today Media Network, March 9. http://indiancountrytodaymedianetwork.com/article/pine-ridge-residents-halt-canadian-mine-equipment-transportation-through-reservation-102246

Schilling, Vincent. 2013a. "Behind the Front Lines of the Elsipogtog Battle over Fracking." Indian Country Media Network, November 20. http://indiancountrytodaymedianetwork.com/2013/11/20/behind-lines-elsipogtog-aptn-journalist-ossie-michelin-gives-backstory-152354

Schilling, Vincent. 2013b. "Woman with Eagle Feather: The Photo 'Heard' round the World." Indian Country Today Media Network, November 21. http://indiancountrytodaymedianetwork.com/2013/11/21/woman-eagle-feather-photo-heard-round-world-152357

Smith, Mitch. 2015. "Grass-Roots Push in the Plains to Block the Keystone Pipeline's Path." *New York Times*, May 6. http://www.nytimes.com/2015/05/06/us/grass-roots-push-in-the-plains-to-block-the-keystone-pipelines-path.html

Sontag, Deborah, and Brent McDonald. 2014. "In North Dakota, a Tale of Oil, Corruption and Death." *New York Times*, December 29. http://www.nytimes.com/2014/12/29/us/in-north-dakota-where-oil-corruption-and-bodies-surface.html

Stone, Laura. 2013. "Chief Atleo Slams RCMP for 'Extreme Use of State Force' in N.B. Shale Gas Protest." Global News (Canada), October 18. http://globalnews.ca/news/910934/chief-atleo-slams-rcmp-for-extreme-use-of-state-force-in-n-b-shale-gas-protest/?utm_source=facebook-twitter&utm_medium=link&utm_campaign=community

"Tar Sands Opposition Unites Burnaby Leaders, First Nations, as 60 Arrested in B.C. Protests." 2014. Indian Country Today Media Network, November 26. http://indiancountrytodaymedianetwork.com/2014/11/26/tar-sands-opposition-unites-burnaby-leaders-first-nations-60-arrested-bc-protests-158031

Taylor, Kevin. 2013. "Nez Perce Leaders Stand Firm on Frontlines of Mega-Load Transport." Indian Country Today Media Network, August 9. http://indiancountrytodaymedianetwork.com/2013/08/09/nez-perce-leaders-stand-firm-frontlines-mega-load-transport-150809

Tollefson, Jeff. 2013. "Climate Science: A Line in the Sands." *Nature* 500 (August 8): 136–37. http://www.nature.com/news/climate-science-a-line-in-the-sands-1.135150

"Tribal Members Sign Treaty Calling for an End to Alberta Oil Sands Development and Keystone XL." 2013. Indian Country Today Media Network, January 31. http://indiancountrytodaymedianetwork.com/2013/01/31/tribal-members-sign-treaty-calling-end-alberta-oil-sands-development-and-keystone-xl

Troian, Martha. 2013a. "First Nation Moves to Evict Fracking Co. from Lands Held in Trust." Indian Country Today Media Network, October 3. http://indiancountrytodaymedianetwork.com/2013/10/03/elsipogtog-first-nation-wrests-lands-fracking-co-mikmaq-treaty-day-151564

Troian, Martha. 2013b. "Police in Riot Gear Tear-Gas and Shoot Mi'kmaq Protesting Gas Exploration in New Brunswick." Indian Country Today Media Network, October 17. http://indiancountrytodaymedianetwork.com/2013/10/17/police-riot-gear-tear-gas-and-shoot-mikmaq-protesting-gas-exploration-new-brunswick

Troian, Martha. 2014. "Court Tells Lubicon Lake Cree in Alberta to Stop Blocking Fracking Activities." Indian Country Today Media Network, February 6. http://indiancountrytodaymedianetwork.com/2014/02/06/court-tells-lubicon-lake-cree-alberta-stop-blocking-fracking-activities-153435

"Turtle Mountain Tribal Council Bans Fracking." 2012. Indian Country Today Media Network, November 27. http://indiancountrytodaymedianetwork.com/article/turtle-mountain-tribal-council-bans-fracking-64866

"Unions Back Lubicons." 2000. Indigenous Environmental Network, August 16. http://www.ienearth.org/lubicon.html#canada

"Welcome to Fearless Summer: Protesters Block Keystone XL Construction." 2013. Indian Country Today Media Network, June 24. http://indiancountrytodaymedianetwork.com/2013/06/28/fracking-troubles-atlantic-first-nations-after-two-dozen-protesters-arrested-150192

Woodard, Stephanie. 2013. "Planned Oil Pipeline Must Cross Pine Ridge's Water-Delivery System." Indian Country Today Media Network, September 21. http://indiancountrytodaymedianetwork.com/article/planned-oil-pipeline-must-cross-pine-ridge%25e2%2580%2599s-water-delivery-system-54942.html.

"Wyoming Groundwater Again Tests Positive for Fracking-Related Chemicals on Wind River Reservation." 2012. Indian Country Today Media Network, September 27. http://indiancountrytodaymedianetwork.com/article/wyoming-groundwater-again-tests-positive-for-fracking-related-chemicals-on-wind-river-reservation-136259

# Chapter 6

# *Mining: Mother Earth or Mother Lode?*

To many Native Americans—those with a naturalistic philosophy—mining is the ultimate insult, a rape of Mother Earth transformed into mother lode, thus the number and intensity of protests by indigenous peoples against mining across Turtle Island, or North America. The mining of uranium (see Chapter 1) is most notably odious; it is such a natural breach that Navajo cosmology warns against it. Mercantile capitalism has no such qualms about mining. It is, in fact, the very basis of a system that survives and thrives by making and selling things, thus the germ of conflict: Native American homelands contain vast stores of exploitable natural resources that corporations find useful and profitable, often available from governments at prices far below market value—or so they were until the "other" started talking back.

The catalog of mining and its toxic legacy on Native American lands is vast. Many Native peoples across North America have organized against mining on their homelands, notably in Alaska, where the world's largest sockeye salmon run could be imperiled by the proposed Pebble Mine, the largest gold and copper strip mine on Earth, which has the potential to produce as much as 80 billion pounds of copper. A permit for that mine was denied in 2014 by the U.S. Environmental Protection Agency (EPA), but mining companies are challenging it.

In Wisconsin, copper, zinc, and sulfide mining are being proposed, and they are being resisted by Native peoples. Silver mining in Mexico is poisoning indigenous children. Coal strip mining is an important issue for Hopis and Navajos, whose shepherds have lost many of their sheep to the toxic plumes of the Four Corners Power Plant.

## *THE QUAPAW TRIBE'S TOXIC MINING MESS*

We begin our toxic journey with the Quapaw in Oklahoma, whose homeland is now nearly uninhabitable. On March 25, 2013, the Quapaw sued the U.S. federal government for $75 million, alleging failure to clean up the nation's largest lead and zinc mine, which had produced billions of tons of ore. Much of the reservation's water had become polluted. Bullfrog Films and director Matt Myers made a documentary film on the Quapaws' pursuit of justice, released in 2011. The film, simply titled *Tar Creek*, provides a searing portrait of human and natural damage wrought on a 47-square-mile area in northeastern Oklahoma now known as the Tar Creek Superfund Site, where residents are still seeking relief from acidic mine water in creeks, lead poisoning of children, and sinkholes that swallow backyards of contaminated houses.

With very little editorial comment—none is required—*Tar Creek* describes the demise of the Native American and European American communities that supplied the materials basic to two world wars. The Quapaws, whose homeland once encompassed much of Arkansas, were removed to a small tract of Indian Territory land during the 1830s; whites later moved in around them. The Quapaws, once a nation of at least 35,000 people, were reduced to about 150, mainly by smallpox and other diseases. They were battling back when the mining ruined their new homeland.

The film's interviews are striking but lack a sense of coherence. Even so, the story is powerful. Following the mining, mountainous piles of "chat" (mining tailings) remained; local waters infused with sulfur ran rust-red. No one told the residents that the enormous piles of chat caused developmental difficulties for the children who played on them. Families laid out picnics on the chat piles, and fathers built sandboxes filled with the gray sandy material that was poisoning their children.

Personal detail is the film's strong point. Environmental problems, we are told, are people problems as *Tar Creek* follows a tale of bureaucratic ineptitude. The federal government initially spent $138 million replacing soil around many houses, an average of $70,000 per house in an area where the homes themselves averaged $58,000 in value. All told, the remediation effort did very little to address the systemic problems caused by leaks from the chat piles and massive water pollution. One report estimated that 50 large trucks would be required around the clock for 40 years to clear the chat piles—and then where would they dump it?

The film follows people's responses as the government conducted a slow-motion buyout in which very few people received anything close to what they would need to start over elsewhere. The camera pans a dead landscape largely bereft of animal, fish, and bird life as human communities slowly disintegrate. One resident compares his community's fate to Rachel Carson's *Silent Spring*. He is wrong, in a sense, because this environmental apocalypse—"forty-seven square miles of prairie turned into permanent wasteland"—is worse than anything described in Carson's book.

By mid-2014, parts of the 40-square-mile Tar Creek Quapaw toxic mining site, most notably areas called the Catholic Forty and Chat Base 11 North, were being cleaned up as part of the first Superfund project by the Quapaw tribal government, cooperating with state and local environmental agencies. "We completed the first cleanup less expensively and better than previous efforts," said Quapaw Chairman John Berrey. "Our goal is to make this land useful and productive again. We live here and we care about the outcomes, so we are very pleased to have these two new agreements in place" ("Quapaw Take Lead" 2014). Roughly 72,000 tons of contaminated material had been removed from the Catholic Forty by 2014—a start with a lifetime of decontamination to come.

## *ALASKA'S PEBBLE MINE VERSUS SALMON RUNS*

The Pebble Partnership proposed a 1,700-foot-deep, two-mile-wide open-pit mine (with underground mines as well) 200 miles southwest of Anchorage that would dump mining wastes into some of the world's most fertile salmon spawning grounds, home of the Earth's largest wild sockeye run. Twenty-five federally recognized Alaskan Native tribes live in the Bristol Bay watershed, having maintained a culture and economy based on salmon for at least 4,000 years. On average, 37 million sockeye return each year to Bristol Bay, supporting more than 12,000 Eskimo and non-Native commercial fishing people. The Pebble Mine could turn part of this salmon run into a toxic dump.

Joby Warrick (2015) of the *Washington Post* described "the region's marshy lowlands [which] are dotted with kettle lakes and crisscrossed by countless streams. In the spawning season, icy rivers turn into roiling torrents of red fish as tens of millions of Pacific salmon make their way inland to spawn." The area has been known as a rich potential source of gold for 30 years, but the day is past when freelancers could pan chunks of gold from streams or dig them out of the earth. The Pebble Mine's riches are embedded in fine particles throughout the soil. "To extract the gold profitably," wrote Warrick, "requires excavating huge amounts of soil and rock from an open-pit mine that could, according to preliminary design plans, eventually cover a seven-square-mile area, making it one of the world's largest. The project would create mountains of rocky spoils, while the wastewater from extracting the ore would be kept in large containment ponds."

Anglo American, a mining conglomerate headquartered in London, proposed to join with British Columbia's Northern Dynasty Minerals in digging the mine near Bristol Bay, with facilities to crush and separate rock as well as tailings (waste) dumps that would eventually become far larger than the mine, itself the largest open-pit mine in North America. Potential yield over the life of the mine has been estimated at 110 million ounces of gold and 80 billion pounds of copper. Even so, the mining companies insisted that salmon runs and wildlife habitat would not be harmed. An EPA assessment issued in March 2012 came to a different conclusion: "The project would entail the destruction of 55 to 87 miles worth of so-far-untouched streams and 2,500 acres of wetlands—and that doesn't even begin to address the potential for disaster if any of the tailings ponds were to leak their acidic water and heavy metals into salmon spawning grounds" ("Bristol Bay" 2012). The EPA assessment was a best case, assuming that earthen dams (some of them 700 feet high) holding tailings ponds intact would not break or leak.

Callan J. Chythlook-Sifsof, a Yupik/Inupiat Eskimo who was raised in Aleknagik, Alaska, an indigenous Yupik Eskimo village near the site of the proposed mine, recalled his youth in a community whose members fished for a living before he became a member of the U.S. snowboarding team in the 2010 Winter Olympics:

> I spent my summers on the back deck of family fishing boats working multiple fisheries. The boats and fish camps are maintained by generations of families harvesting salmon not only for income, but also for food. I remember long days of processing hundreds of pounds of salmon, setting nets, cleaning and filleting, filling tubs of salt brine, putting fresh water in clean white buckets and hanging neat rows to dry and smoke. . . . As a child, I had no idea what magic this life was—it was just the way we did things. It's the way many Alaska Natives live—through self-reliance and hard work to harvest the many gifts of the land and sea.
>
> This subsistence way of life that is thousands of years old is threatened by the plans of a British and Canadian mining partnership to dig a huge mine in the heart of our productive, healthy watershed. Risking our lands, waters and fish for a short-term mega-mine like Pebble is a terrible idea. Eighty-one per cent of Native shareholders in the Bristol Bay Native Corporation—composed of more than 9,000 Native Alaskans with ancestral ties to the Bristol Bay region—opposed the mine in a 2011 survey. And that's despite the promises of jobs and continuing efforts by the Pebble Partnership . . . to buy support through grants and giveaways to communities and hundreds of millions spent to develop the mine. (Chythlook-Sifsof 2013)

In a survey by the Natural Resources Defense Council, 68 percent of Alaskans opposed the mine. More than 75 percent of 900,000 public comments on the project collected by the EPA opposed the project. "It's truly alarming," wrote Chythlook-Sifsof (2013), "when Pebble's chief executive officer, John Shively, blithely says that, sure, the mine will damage some salmon habitat, but the company will just build 'comparable' habitat nearby. His comments show a lack of understanding of salmon life cycles, habitat and ecosystems—not to mention the people of Bristol Bay." Native peoples urged the EPA to reject the plans for the mine under the aegis of the federal Clean Water Act.

A total of 360 scientists from the United States, Poland, Canada, Australia, and other countries signed a letter on February 3, 2014, urging the EPA to disapprove the Pebble Mine development. "Over three years, EPA compiled the best, most current science on the Bristol Bay watershed to understand how large-scale mining could impact salmon and water in this unique area of unparalleled natural resources," said Dennis McLerran, regional administrator for EPA Region 10. "Our report concludes that large-scale mining poses risks to salmon and the tribal communities that have depended on them for thousands of years. Based on the results of the assessment, we are very concerned about the prospect of large-scale mining in the unique and biologically rich watersheds of southwest Alaska's Bristol Bay," the scientists wrote. "The preponderance of evidence presented in the watershed assessment indicates that large-scale hard rock mining in the Bristol Bay watershed threatens a world-class fishery and uniquely rich ecosystem, and we urge the Administration to act quickly to protect the area. Therefore, we urge EPA to use its authority under the Clean Water Act to take the necessary next steps to protect Bristol Bay" ("360 Scientists" 2014).

In September 2013, Anglo American withdrew from an industrial consortium seeking to develop the Pebble Mine. Anglo American had been half owner with a mineral exploration and development company, Northern Dynasty, based in Vancouver, British Columbia. "Our focus has been to prioritize capital to projects with the highest value and lowest risks within our portfolio, and reduce the capital required to sustain such projects during the pre-approval phases of development as part of a more effective, value-driven capital allocation model," said Mark Cutifani, chief executive of Anglo American ("Anglo American Withdraws" 2013).

Following the appeal by the 360 scientists as well as a visit to Washington, D.C., by representatives of the 31 Native villages in the Bristol Bay watershed—all adamantly opposed to the Pebble mine proposal—in late February 2014, the EPA refused to allow the U.S. Army Corps of Engineers to issue a permit to Northern Dynasty Minerals Ltd of Canada. "Extensive scientific study has given us ample reason to believe that the Pebble Mine would likely have significant and irreversible negative impacts on the Bristol Bay watershed and its abundant salmon fisheries," EPA Administrator Gina McCarthy said. "It's why E.P.A. is taking this step forward in our effort to ensure protection for the world's most productive salmon fishery from the risks it faces from what could be one of the largest open pit mines on earth. This process is not something the Agency does very often, but Bristol Bay is an extraordinary and unique resource." The EPA statement pointed out that water quality is crucial to the area's average annual run of 37.5 million sockeye salmon as well as smaller runs of Chinook, coho, pink, and chum salmon. "In addition, it is home to more than 20 other fish species, 190 bird species, and more than 40 terrestrial mammal species, including bears, moose, and caribou," the EPA said. Digging an open-pit gold and copper mine a mile deep and 2.5 miles wide would require at least three earthen tailings dams as high as 650 feet that "would irreparably harm the ecosystem" ("Native Alaskans" 2014).

On December 16, 2014, President Obama signed a presidential memorandum that declared the Bristol Bay watershed, "these beautiful and pristine waters," off limits from all future oil and gas drilling ("Obama Declares" 2014). He invoked Native Alaskans' key role in preserving the watershed by shielding it—until, of course, some future Republican president finds a way to nullify the order. The order did not address the Pebble Mine proposal, which had already been denied. With so much at stake, however, the miners and drillers will not take "no" for an answer. A reversal could be no further away than a Republican victory in the next U.S. presidential election.

## *"POUR HIM A TALL ONE!"*

Before dawn on August 4, 2014, three weeks after the Pebble Mine's rejection and as its sponsors were getting ready to resubmit the proposal, a tailings pond burst in British Columbia's Mount Polley Mine, sending 4 billion gallons of mining waste (2.6 billion gallons of water mixed into a slurry with 1.2 billion gallons of fine sediment laced with metals) into rivers and lakes awaiting return of salmon. Knight Piesold Consulting, the firm that designed the burst dam, had also been hired as a contractor to design the tailings pond for the Pebble Mine, according to a report filed in 2006 by Northern Dynasty Mines Inc. with the Alaska Department of Natural Resources.

Critics of the Pebble Mine proposal saw the Mount Polley Mine tailings spill as a prelude and a warning. "We don't want this to happen in Bristol Bay," said Kim Williams of Dillingham, director of Nunamta Alukestai (which means Caretakers of Our Land), a conservation group of Alaska Native tribes and corporations, to *The Cordova Times*. "With all the similarities between Pebble and the Mount Polley copper mine, we're urging the EPA to take immediate action to finalize mine waste restrictions in Bristol Bay. Our hearts go out to those in British Columbia who live downstream from this devastating mine failure." "It's Bristol Bay's worst nightmare," said Carol Ann Woody, a fisheries scientist with the Center for Science in Public Participation ("Company" 2014).

"The Pebble project would be bigger—a lot bigger," wrote Joel Reynolds, western director and senior attorney for the National Resources Defense Council, in the *Huffington Post*. "While Imperial Metals has been mining about 20,000 tons per day at its Mount Polley mine, Northern Dynasty has anticipated about ten times that at Pebble, with a tailings pond many times larger in footprint and scale" ("Company" 2014).

After Imperial Metals President Brian Kynoch bragged that he would drink the water downstream after the Mount Polley spill, the Indian Country Today Media Network headlined, "Pour Him a Tall One!" "It's very close to drinking water quality, the water in our tailings," Kynoch said, according to the Canadian Broadcasting Corporation news. "There's almost everything in it but at low levels. . . . No mercury, very low arsenic and very low other metals. . . . I apologize for what happened. If you [had] asked me two weeks ago if this could have happened, I would have said it couldn't" ("Pour" 2014, 2–14).

Several First Nations people live in the area of the spill and depend on its salmon runs. "Our communities are filled with sorrow, frustration and anger as they are left wondering just what poisons are in the water, and what is being done to address this disaster," said Williams Lake Chief Ann Louie and Xat'sull Chief Bev Sellars. "Monday's devastating tailings pond breach is something that both our First Nations have lived in fear of for many years," the chiefs said. "We have raised repeated concerns about the safety and security of this mine, but they were ignored. Now we are being ignored again. Enough is enough" ("Horrific" 2014).

NASA's Earth Observatory described the Polley Mine dam breach as viewed from satellites high above the Earth:

> On August 5, nearly all of the wastewater in the retention basin had drained, exposing the silty bottom. Hazeltine Creek, normally about 1 meter (3.3 feet)

> wide, swelled to a width of 150 meters (490 feet) as a result of the spill. In the aftermath of the flood, a layer of brown sediment coated forests and stream valleys affected by the spill. Notice how much forest immediately north of the retention basin was leveled. Debris, mainly downed trees, are visible floating on Quesnel Lake. (Note that some of the variations in colors between the two images are caused by different lighting conditions and viewing angles.) The breach released more than 10 million cubic meters (350 million cubic feet) of water and 4.5 million cubic meters (150 million cubic feet) of sand into Polley and Quesnel Lake, according to the British Columbia's Ministry of Environment. That is enough water to fill 4,000 Olympic-sized pools. ("Dam Breach" 2014)

Native peoples were not persuaded by Imperial Metals' insistence that the mine waste behind their ruined dam was not harmful. "As of last night, Department of Fisheries and Oceans has banned salmon fishing in the Cariboo and Quesnel Rivers due to Mount Polley," said Chief Bob Chamberlin, vice president of the Union of B.C. Indian Chiefs, on August 6. "Mount Polley will have an immediate and devastating effect on First Nations like Lhtako Dene, Lhoosk'uz Dene, Nazko, and Esdilagh who may not be able to fish for salmon at all this year. First Nations are anxiously awaiting the water-test results, the possible . . . closures afterwards and the harmful impacts on future salmon runs of the Fraser" ("Horrific" 2014).

## *MINING'S "NEW PROSPERITY" = NATIVE DESTRUCTION*

To the Taseko mining company, a massive strip mine for gold and copper meant "new prosperity," but to the Tsilhqot'in (Chilcotin) indigenous people of British Columbia, it meant genocide. They united with environmentalists and other indigenous peoples across Canada and won a reprieve. "As a Nation, we've made a decision to focus our energies on the real battle of defeating this project, full-stop," said Chief Marilyn Baptiste of the Xeni Gwet'in, one of the communities making up the Tsilhqot'in National Government ("Tsilhqot'in" 2012).

Over several years, the Tsilhqot'in and their allies forced the scaling down (and finally cancellation) of the mine. An early proposal would have severely damaged the trout fishery in Teztan Biny (Fish Lake). The new plan said it would spare that area but still would have turned a smaller body of water, Little Fish Lake, into a tailings pond. To stop it, they went to court and set up a blockade. "It's a simple No," said Marilyn Baptiste, chief of Xeni Gwet'in, the Tsilhqot'in band nearest to the mine site. "Our land is not for sale. Our position has not changed and cannot change for the destruction of our lands, our waters and our way of life. Our wild rainbow trout has survived in that lake system for hundreds of years, as our people have. We will not, and cannot, agree to such destruction in the headwaters by Taseko" (Ball 2012).

The company asserted that its plans were environmentally responsible, but the opponents did not relent. In June 2012, Taseko delivered a revised proposal at its annual meeting in Vancouver that mitigates environmental damage, according to President and Chief Executive Officer Russell Hallbauer. "This project . . . holds exciting potential for the company's stakeholders, including shareholders and local communities," Hallbauer wrote. "These revised plans address the environmental concerns identified in the original environmental assessment process, and importantly, includes the preservation of Fish Lake" (Ball 2012).

Chief Joe Alphonse, tribal chairman of the Tsilhqot'in National Government, reiterated his nation's opposition to the mine. "We're not opposed to development," Alphonse said, "but this is not the way to do it. . . . This is the first time in Canada that a mine has been approved a second time [after being rejected]. . . . They've reloaded and they're coming back with a Conservative majority federally." Alphonse described Teztan Biny (Fish Lake) as a thriving ecosystem of vital importance. "You can see the fish jumping out of Fish Lake, and wolf tracks, moose tracks," he said. "We have the most consistent sockeye run in North America—the only run on the Fraser [River] that's still healthy" (Ball 2012).

The Union of B.C. Indian Chiefs, representing British Columbia's First Nations, also voiced support. Its president, Grand Chief Stewart Phillip, accused Prime Minister Stephen Harper's Conservative government of colluding with mining companies. "We were shocked and appalled when we learned the Harper government and his cronies in the mining industry got a second kick at the can," Phillip said. "The second proposal . . . is far more destructive" (Ball 2012).

On February 26, 2014, the Canadian Ministry of the Environment issued a final rejection of the New Prosperity Mine as the Tsilhquot'in and their allies celebrated across the country following two decades of conflict. This was the government's third rejection, centered on the company's proposal to turn Fish Lake, which is scared to the Tsilhqot'in, into a tailings pond. The third proposal made a case that the mine and area lakes could coexist. Neither the Native people, their non-Indian allies, nor the Ministry of the Environment was buying it. The government asserted that the mine, which the company said could be one of the largest new sources of gold and copper on Earth, was "likely to cause significant adverse environmental effects that cannot be mitigated" (Lazenby 2014).

"We are celebrating this decision to reject once again this terrible project, which threatened our pristine waters, fish and aboriginal rights," said Chief Joe Alphonse, tribal chair for the Tsilhqot'in National Government. "We commend the federal government for not bowing to industry lobbying and instead respecting the science and the independent process which came to the conclusion that this project would have devastating impacts on the environment and our Nation's ability to practice our rights in a sacred spiritual site. These impacts could not be mitigated" (Smitten/Vimeo 2014).

Taseko may return, however, after having already committed $300 million to the mine. The government said that a new proposal was possible if environmental concerns were addressed. Russell Hallbauer, president and chief executive officer of Taseko, said that New Prosperity "could be built to a high standard of environmental integrity, including the full protection of Fish Lake" (Lazenby 2014).

Shortly after the environmental ruling favoring the Tsilhqot'in's 3,000 people in six communities, on June 26, 2014, the Supreme Court of Canada granted aboriginal title to more than 1,750 square kilometers (about 675 square miles). In Canada, this was the first grant of aboriginal title by the federal Supreme Court, a watershed legal ruling. This ruling overturned a Court of Appeal decision in 2012 that Tsilhqot'in bands were "seminomadic" with few permanent encampments to allow for aboriginal title. "There is no suggestion in the jurisprudence or scholarship that Aboriginal title is confined to specific village sites or farms, as the court of appeal held," the Supreme Court ruling said. "Rather, a culturally sensitive approach suggests that regular use of territories for hunting, fishing, trapping and foraging is 'sufficient' use to ground Aboriginal title."

### *COAL: PACIFIC NORTHWEST TRIBES PROTEST TRANSPORT*

For fossil-fuel firms in North America with an eye on burgeoning markets in China and India, the shortest line between profit and the coal fields of Wyoming, the oil deposits of the Bakken formation in the Dakotas, and the tar sands oil of Alberta may be the Pacific Northwest, most notably the "Salish Sea," as some Native people call it, the strip of bays, straits, and inlets between Vancouver, British Columbia, and Seattle, Washington. By 2014, Native peoples in that area, allied with several environmental groups, seemed to have fought plans to ship coal to the Pacific Coast and Columbia to a standstill. What remained was Cherry Point, Washington, in very northernmost Puget Sound, a few miles south of the Canadian border, already the site of a large oil-refining complex.

Plans for a terminal at the Port of Grays Harbor, Washington, near the Quinault reservation, were shelved in 2012. In Oregon, three proposals for coal-export terminals met less-than-lukewarm receptions. Morrow Pacific in St. Helens also ran into Native American and environmental resistance; plans for a coal terminal at Port Westward were abandoned by Kinder Morgan in 2013, and Project Mainstay, proposed for the Port of Coos Bay, had lost most of its investors by 2014. Yet another one, the Millennium Bulk Terminal in Longview, Washington, was still being pursued.

Cherry Point is also a favored fishing ground of the Lummis, whose reservation lies north of Bellingham and south of Vancouver and who adamantly oppose plans for the Gateway Pacific Terminal, meant to export 48 million tons of coal from eastern Montana and Wyoming's Powder River basin, averaging nine coal trains a day. They see the terminal as a long-term threat to their fishing, a 2,500-year-old tradition, a multi-million-dollar annual harvest of salmon, carp, and halibut, basic to their *schelangen* ("how life ought to be lived").

Pacific Terminals, a subsidiary of SSA Marine, is among the largest shippers of coal on Earth. Amber Cortes, writing in *Grist*, outlined the set-piece battle shaping up:

> At Cherry Point, the site of the proposed coal terminal, the water is great for fishing: the depth of the sea bottom drops off quickly from the shore and

> attracts fish like surf smelt, sand lance, and herring that come to feed and spawn. It's one of the reasons Cherry Point is so special, and a state aquatic reserve. That very deep water, so close to shore, is what makes Cherry Point so special to Pacific Terminals and SSA Marine, too. You can't bring tankers and bulk carriers that draft in excess of 50 feet deep into just any old port. "But what's good for the shipper is critical for the fish," says Fred Felleman, a marine biologist who consults with Pacific Northwest tribes and environmental groups about shipping impacts on the marine environment. (Cortes n.d.)

Many people in the area who had expressed concern that coal shipments threaten the area's natural bounty recalled a coal-ship spill in 2012 in Vancouver at Westshore Terminals. The ships that carry coal are larger than oil tankers and tougher to steer though rock-studded inlets. They also are not legally required to employ tugboats as escorts. The upshot, according to the Indian–environmentalist alliance, is a risky future on the Salish Sea, a complex marine environment full of islands and dense with other shipping traffic.

Pacific Terminals made no friends in June 2011, when it bulldozed wetlands at Cherry Point without environmental permits. The same area was registered as a state archaeological site containing 3,000-year-old Lummi human remains (Cortes n.d.). Soon the Lummi were burning symbolic checks in front of television cameras with this message: no amount of coal money would buy them off. Totem poles were fashioned by master carver Jewell James in protest of the coal terminal, to be carried cross-country, starting at the desecrated earth of Cherry Point:

> The totem pole will stand at 19 feet tall when completed. Carved into the pole is Mother Earth, lifting a child up to receive the teachings of the heavens. On the bottom of her dress are four warriors who are guardians of the sacred earth-power, and a snake, representing the power of the earth traveling upwards. (Cortes n.d.)

James said that the totem pole "depicts a woman representing Mother Earth, lifting a child up; four warriors, representing protectors of the environment; and a snake, representing the power of the Earth" (Walker 2014). It was carried to the source of the coal as well as the tar sands. The Lummis, in the meantime, assert that the coal terminal, if built, will impede their rights under the Treaty of Point Elliott (1855), as upheld in U.S. federal courts during the 1970s. This argument already has been used to persuade the Army Corps of Engineers to deny permission to build a commercial salmon farm in the same area. Protection of Quinault fishing rights was a major factor in the defeat of two proposed oil terminals at Grays Harbor.

A protest of coal transport to the Gateway Pacific Coal terminal assembled at Cherry Point (which the Lummi call *Xwe'chi'eXen*), led by Native fishing people with non-Indian allies. "We have to say 'no' to the coal terminal project," said Cliff Cultee, chairman of the Lummi Nation. "It is our *Xw' xalh Xechnging* [sacred duty] to preserve and protect all of *Xwe'chi'eXen* [the Lummi homeland]" ("Northwest Tribes Step Up" 2012). The Indian Country Today Media Network reported that "during the protest, the tribes held a ceremony of thankfulness, remembrance and unity to solidify their commitment." The statement noted that Lummi Indians have the largest Native fishing fleet in the United States and have fished off Cherry Point for thousands of years.

Writing on August 11, 2014, in the *Bellingham Herald*, Jewell James of the Lummi Nation's Sovereignty and Treaty Protection Office asserted that the coal terminal proposed at Cherry Point poses "a tremendous ecological, cultural and socio-economic threat" to Pacific Northwest indigenous peoples. "We wonder how Salish Sea fisheries, already impacted by decades of pollution and global warming, will respond to the toxic runoff from the water used for coal piles stored on site." James continued, "What will happen to the region's air quality as coal trains bring dust and increase diesel pollution? And of course, any coal burned overseas will come home to our state as mercury pollution in our fish, adding to the perils of climate change" (James 2014). Marianne Maumus of Ochsner Health Systems told the *New Orleans Times-Picayune* that coal dust contains heavy metals, including arsenic, cadmium, and mercury, and can cause cancer, neurological, renal, and brain development problems. "I think the risk is real. I think there is a lot of potential harm from multiple sources," Maumus said (Walker 2014).

The Affiliated Tribes of Northwest Indians passed a resolution in September 2012 requesting a full environmental analysis by the Army Corps of Engineers of several coal-transport proposals for terminals on the Columbia River's Port of Morrow and Longview as well as at Cherry Point north of Bellingham near the Lummi Nation (near an existing oil refinery). "This is not about jobs versus the environment," said James in a Lummi Nation statement after the protests. "It is about what type of jobs are best for the people and the environment" ("Northwest Tribes Step Up" 2012). Despite the Lummis' strident opposition to this project, its sponsors have continued to knock on their doors through at least 2015 (when this book went into production), as Lummi Nation Chairman Tim Ballew pointed to a number of accidents involving coal and oil–carrying rail cars across North America during the previous several years and said:

> I can assure you that we have carefully considered the impacts associated with this project and have concluded that these impacts simply cannot be avoided, minimized or mitigated. . . . While we appreciate your desire to engage on these issues, we remain steadfastly opposed to this project and do not see the utility in pursuing any further discussion. Please be advised that we fully intend to ensure that our treaty rights are honored and that the spiritual and cultural significance of Xwe'chi'cXen is protected. (Walker 2014)

In August 2014, the Oregon Department of State Lands rejected a coal-terminal application along the Columbia River, citing Native treaty and fishing rights considerations. Local tribes celebrated, and others to the north, in Washington State, saw the decision as a possible precedent setter for fossil-fuel terminals that they were resisting. The department said that "while the proposed project has independent utility, it is not consistent with the protection, conservation and best use of the state's water resources" and that the applicant did not provide sufficient analysis of alternatives that would avoid construction of a new dock and impacts on tribal fisheries. Ambre Energy, which had applied to ship 8.8 million tons of coal a year to Asia via the Port of Morrow's Coyote Island Terminal, has said that it would appeal.

"Today's landmark decision reflects what is in the best interest of the region, not a company's pocketbook," said Carlos Smith, chairman of the Columbia River Inter-Tribal Fish Commission. "This decision is one that we can all celebrate. It reaffirms the tribal treaty right to fish, is in the best interest of the Columbia Basin's salmon populations, and our communities. It is a reflection of what is best for those who would be forced to live with the consequences of Ambre's proposal, not what is best for those who would profit from it. This is the beginning of the end for this toxic threat—the Tribes will stand with the State to protect its sound decision" ("Treaty Victory" 2014).

"The State's action makes a strong policy statement by recognizing Tribal Sovereignty and the Treaty Rights of the Columbia River tribes," said Lummi Nation Chairman Timothy Ballew II. "Such decisions are few and far between. This is important not just for the Yakama and Umatilla but all Indian fishing tribes. Together we can, and will, protect our way of life." Yakama Tribal Council Chairman JoDe Goudy, through whose territory trains would pass en route to a terminal on the Columbia River, said, that the "Yakama Nation will not rest until the entire regional threat posed by the coal industry to our ancestral lands and waters is eradicated" ("Treaty Victory" 2014).

Native leaders in the Pacific Northwest have continued to rally against plans to transport coal from the Pacific Northwest to Asia through waterways whose fish and shellfish sustain their lives, asserting, according to the Indian Country Today Media Network, that "treaty fishing rights are meaningless if there are no healthy fish populations left to harvest" (Hansen 2012). Since the Boldt decision (1974), which allocated up to half the salmon in much of western Washington to Native fishing people, Native peoples have worked with state authorities to sustain the runs. Coal transport threatens that effort. Coal dust from trains may degrade human health as well as that of wildlife and fish. The companies are planning to add rail capacity to the single line that was transporting coal over the Cascades and Rockies from Wyoming's and Montana's Powder River mining country. Trains carrying coal from Wyoming reach a mile in length, with plans to double that capacity.

Lummi Indian Business Council Chair Tim Ballew wrote to the Army Corps of Engineers on July 30, 2012, "Any impact on the Lummi treaty fishing right is inherently an impact on the Lummi way of life. . . . We believe that the Corps should see that these projects would without question result in significant and unavoidable impacts and damage to our treaty rights" ("Lummi Nation" 2012). The letter said that the Lummi Indians "maintain the largest Native fishing fleet in the United States. Moreover, Lummi fishers have worked in the *XweChiexen* [Lummi homeland's] fishery for thousands of years. . . . The Gateway Pacific export terminal would be the largest such terminal on Turtle Island's [North America's] west coast. . . . It would significantly degrade an already fragile and vulnerable crab, herring and salmon fishery, dealing a devastating blow to the economy of the fisher community."

The *PBS NewsHour* took up the Lummis' resistance to the Cherry Point coal terminal proposal on August 2, 2012, during which health professionals described the dangers of coal dust and diesel exhaust as Lummi fishing people spoke of threats to their harvest. As proposed by SSA Marine in 2012, the coal terminal would export 54 million metric tons per year of "bulk commodities," including 48 million metric tons of coal. "If the Lummis come to that position, it will make us reassess the direction we are going," said Muffy Walker, the head of the Army Corps of Engineers' regulatory branch in Seattle. "We have denied permits in the past, based on tribal concerns" ("PBS News Hour" 2012).

Opponents of the proposed coal-export terminals decry the noise, dirt, and odor of the trains and say that coal dust may threaten herring, Chinook salmon, and killer whales. Environmental problems already have reduced salmon runs to the lowest since Judge George H. Boldt's landmark decision, according to a report from the Northwest Indian Fisheries Commission. The Indian Country Today Media Network reported, "Many worry that such a network of coal transport would degrade salmon and cultural-foods habitats as well as affect treaty rights. Among the concerned groups are the Columbia River Inter-Tribal Fish Commission (CRITFC), the National Wildlife Federation (NWF)'s Tribal Lands Program, and tribal nations like the Lummi in northwest Washington and the Yakama in eastern Washington" (Hansen 2012).

"Our treaty rights to first foods—salmon, deer, chokecherries, mountain sheep, wild roots and other resources—are retained by the tribes, and they weren't given to us by the government," said Bruce Jim, a Confederated Tribes of Warm Springs elder who serves on the tribe's Fish and Wildlife Committee. "Moving coal through our land will threaten these rights and our connections to these plants and animals" (Hansen 2012). On the coast, a report by the NWF, "The True Cost of Coal," anticipates "habitat degradation from port expansions and shipping traffic; decreased water quality from coal dust; mercury deposits from coal burning and wind-driven transport; increased carbon pollution; and ocean acidification" (Hansen 2012).

Peabody and other coal companies have been planning to ship coal, as much as 150 million of tons a year, by river, rail, and road, mainly to China. By 2013, coal consumption and prices in the United States were in decline. Electrical utilities in the United States used 18 percent less coal in the first half of 2012 than during the first half of 2011. Coal consumption in 2012 was 27 percent below that of its peak year, 2008, following several years of steady growth. Coal is being replaced for electrical power generation by natural gas from fracking, which produces about half the carbon dioxide per energy unit as coal—it's still a fossil fuel but a cleaner one. Coal is now being mined for export to Asia (mainly to China), requiring West Coast terminals in western Washington and along the Columbia River, where Indian tribes have treaty rights for salmon.

"The idea of ... new coal export terminals in Western Washington and Oregon—and the hundreds of trains and barges running from Montana and Wyoming every day to deliver that coal—would threaten our environment and quality of life like nothing we have seen before," said Billy Frank Jr., chairman of the Northwest Indian Fish Commission. "Coal may be a cheap source of energy for other countries, but these export facilities and increased train traffic would come at a great cost to our health, natural resources and communities." Swinomish Tribal Chairman Brian Cladoosby said, "If a coal train or tanker were to spill on the route or in the river at Port Morrow in Oregon, the water ways will carry the pollution throughout the Northwest, and coal dust will be carried through the mountains in the air we all breathe" (Trahant 2012).

The Native peoples pointed out that the coal-export business is being subsidized by the U.S. federal government. The Institute for Energy Economics and Financial Analysis reported in 2012 that coal mined on lands owned by that government is so cheap (less than $10 a ton) that taxpayers are losing $1 billion a year in royalties, as the companies are able to sell what they mine for as much as 10 times what they have paid. "Taxpayers are likely losing out so that coal companies can reap a windfall and export that coal overseas, where it is burned, worsening climate change," said U.S. Senator Edward J. Markey (D–MA) (Riordan 2014, A-23).

Carbon dioxide emitted by burning coal adds to the global burden of greenhouse gases that provoke global warming. According to one analysis, "Projected exports from Cherry Point alone could result in over 100 million metric tons of carbon dioxide annually. The gas has already begun to boost the acidity of near-shore waters, threatening Washington's shellfish industry. The billions of tons of coal burned in Asia every year contribute markedly to global warming. Should the United States be selling them subsidized coal and encouraging this impending disaster?" (Riordan 2014, A-23).

Residents of the Northern Cheyenne reservation in eastern Montana also objected to coal transport, asserting that "coal mining and the construction of a railroad to transport the coal would devastate their essential Otter Creek and the Powder River Basin." The Northern Cheyenne allied with ranchers, hunters, and conservation groups, such as the NWF, against the new coal railroad "and its ruinous effects on its multitude of wildlife, cougars, black bears, elk, mule deer, white-tailed deer and pronghorns all roam the forests and sagebrush prairies that make up this portion of the Powder River Basin in the easternmost portions of Montana and Wyoming. Otter Creek, which supports more than 20 fish species, empties into the Tongue River and eventually the Yellowstone River" (McNeel 2012).

Local residents vented opposition in five public hearings during November 2012. "Almost 100 percent of the people spoke out against it," said Alexis Bonogofsky, a

Northern Cheyenne tribal lands program manager for the NWF in that region. "It was conservative ranchers and liberal ranchers. There were cowboys and Indians. Everyone was just saying, 'We don't want this train.' It was remarkable because in some of those communities, like Forsyth and Ashland, the perception before this was that they supported it" (McNeel 2012).

On December 4, 2012, another hearing was attended by about 800 people in Spokane, Washington. In Seattle, about 2,000 people attended a similar hearing on December 13, 2012, including several Northern Cheyenne who drove overnight from their reservation in eastern Washington. The crowd in Spokane included supporters of the trains and terminals, such as railroad workers, but about 75 percent of the people who spoke were opposed, citing "dangers posed by railroad crossings, slower emergency-vehicle response time, potential coal-dust-related health problems and other concerns. Some came all the way from southeastern Montana, including nine people from the Northern Cheyenne Reservation who made the nearly 1,300-mile round-trip" (McNeel 2012).

"I'm transitioning into something that is a better vision for the future, more involved in renewable energies that should be embraced that we, as global citizens, should embrace," said Jeff King of the Northern Cheyenne, who had been working in the coal industry. "I would like to see our nation embracing renewable energy." Burdette Birdinground, a Crow with Northern Cheyenne relatives, drove from the Flathead reservation. "It will pollute the waters, the air, will ruin the cultural sites in the valley and the hunting grounds," he said. Vanessa Braided Hair said, "stop the destruction of our Mother Earth" (McNeel 2012).

## *THE MOAPA PAIUTE: GOOD-BYE TOXIC ASH: SOLAR IN, COAL POWER OUT*

Quite a few people who would like to retire smoke-belching, sulfur-spewing, carbon-pumping coal-fired power plants might consider taking lessons from the Moapa Paiutes, a tiny American Indian tribe who live a few miles, as the toxic ash flies, north of Las Vegas, Nevada. The Moapa Paiutes, after decades of suffering the suffocating ash clouds of an old coal-fired plant, not only retired it but also convinced the U.S. Department of the Interior to site a solar-powered array nearby, creating jobs.

In 2013, the Reid Gardner Generating Station, a coal-fired power plant adjacent to the Moapa River reservation (about 50 miles north of Las Vegas), was slated for closure in 2017. The closure was good news, but the Moapa Band of Paiutes had a new problem: ramping up pressure on its owner, NV Energy, Inc., to clean up residual pollution. On August 8, 2013, the band and the Sierra Club filed suit in U.S. district court in Las Vegas to legally compel the cleanup. MV Energy was bought out in May 2013 by Warren Buffet's MidAmerican Energy Holdings, which has taken an active role in raising Iowa's use of wind power for electrical generation in Iowa to one-half of its production by 2015. Reid Gardner is the last coal-burning power plant in Nevada. The lawsuit claimed "that the federal Resource Conservation and Recovery Act and the Clean Water

Act have both been violated over the years by dumping that has compromised the health of nearby residents and threatens the drinking water of millions" ("Moapa Paiute Sue" 2013). The lawsuit alleges that for several years, the power plant illegally dumped toxins into the Muddy River, flowing into the Lake Mead reservoir behind Hoover Dam, a source of drinking water for more than 2 million people.

"We are all looking forward to the retirement of the Reid Gardner coal-fired plant that has for decades polluted our reservation," said Vickie Simmons, a leader of the Moapa Band of Paiutes Committee for Health and the Environment. "And for the sake of our families' health, we must ensure that the toxic waste from the power plant is fully cleaned up. The safety of our community and the future of our children depend on it" ("Moapa Paiute Sue" 2013). "Now, we have to find out what kind of remediation they're going to do—a complete restoration, a conversion to gas or some other type of project," Tribal President William Anderson told the Associated Press. "To us, the ultimate goal would be to remove everything and put the land back the way it was. We'll be able to come to come closure after almost 50 years" ("Moapa Paiute Sue" 2013). On Earth Day 2012, the Moapa Paiute had protested the Reid Gardner Power Plant in Las Vegas with members of the Sierra Club members on a 50-mile "Cultural Healing Walk" that took three days in 100-degree temperatures.

Reservation residents said that the plant has been a threat to their health for several years. William Anderson, Moapa Band chairman, said, "The high percentage of thyroid and respiratory problems is a big concern for the tribal members on the reservation. . . . We need a proper study from air monitoring equipment installed on the reservation to study the emissions we're breathing in. That would help determine what needs to be do for our people's health. We also need more stringent storage conditions for coal ash and a study to be conducted to show the health risks associated with breathing in coal ash" ("Moapa Paiute March" 2012).

"We asked the EPA to at least study the plant or give us a grant to do the study to see exactly what poisons were coming from there, but we were told we didn't qualify for the study or the grant," said Vernon Lee, a Moapa Paiute. "In order to get a federal study grant, there have to be 500 people affected. There are only about 300 members of the Moapa Band and not all of us live on the reservation."

The Indian Country Today Media Network reported that

> air pollution takes the form of toxic coal dust, which tribal members say arrives in giant clouds that send people scurrying indoors, but that's not the only problem they have observed. "There are also several settling ponds for coal ash residue, there are enormous piles of coal that are uncovered, and a huge coal ash landfill that is also uncovered," said Barb Boyle of the Sierra Club, in an article published on the *Huffington Post*'s website, written by Mary Ann Hitt, director of the Sierra Club's Beyond Coal campaign. Boyle said that the Tribe "has borne this burden for decades. It's time to stop." The plant was built in 1965, and the Moapa Paiute say they have witnessed their own standard of living plummet over time. "In my era, we were all healthy people," Aletha Tom, who runs the

> Moapa school house, [said]. "We didn't have the asthma, thyroid problems, cancer, diabetes, but we have that on our reservation. It's so major now."

Soon, due to the tribe's persistence, local television stations and newspapers brought the plague of ash-borne illness to a broader audience. A video, *An Ill Wind: The Secret Threat of Coal Ash*, reached people from the tribe's website, reporting that the power station "dumps toxin-laden coal ash, a byproduct of combustion, into landfills that lie just a few hundred yards from the reservation. On windy days, coal-ash dust from the plant billows over the reservation, with clouds so thick that you can see and taste them, tribal members say. At such times, residents don't dare let their children play outside. That apparently offers limited protection, though, as the dust seeps into homes, schools and cars" (Woodard 2012). The Sierra Club, Earthjustice, and Greenpeace assisted.

In their long campaign, the Moapa Paiutes won allies. Senator Harry Reid (D–NV) called the Reid Gardner Power Station a "dirty relic" and supported its closure. In June, Tribal Chairman William Anderson told Indian Country Today Media Network that "the Interior Department gave the Moapa Paiutes fast-tracked approval to build the first-ever utility-scale solar-energy project on tribal lands—which seems especially suitable in this region, with its many months of scorching sun. The 350-megawatt plant, to be built on Moapa Paiute trust land, should generate enough power for 100,000 homes, according to the agency, which says the project will mean lease income for the tribe, as well as new jobs. We should have about 400 jobs at peak construction and 15 to 20 permanent jobs—real career jobs tribal members can look forward to," said Anderson. "Our energy customers will likely be in California, where people have an interest in renewable-energy sources." Construction was planned over several years, along with creation of a preserve for the endangered species desert tortoise. "We're right by an energy corridor, with many above- and below-ground lines for electricity, natural gas, fiber optics, and more easily accessible," Anderson said. "Connecting is much less expensive than if we were farther away" (Woodard 2012).

As bright as the future may be, the Moapa Paiute in 2014 were still suffering the coal plant's effects. "Every home has someone—or even everyone—using a breathing apparatus or inhaler," Anderson said. "We see frequent deaths, the most recent being someone in the home closest to the plant." All along, NV Energy said it was continuing its commitment "to operate the Reid Gardner Generating Station in an environmentally responsible manner, in compliance with all federal and state laws, and in the best interests of its customers" (Woodard 2012).

In May 2014, the Moapa were awarded a second solar-energy array by the Department of the Interior just as constriction began on the first. The second array, on 850 acres of Moapa land, is sited 20 miles northeast of Las Vegas; the first project is about 30 miles due north of Las Vegas. The second array, which will provide power for 60,000 homes, is being constructed to minimize water usage. Secretary of the Interior Sally Jewell said that the array "reflects the Obama administration's steadfast commitment to work with Indian Country leaders to promote strong, prosperous and resilient tribal economies and communities.... This solar project and these grants also deliver on the President's Climate Action Plan goals to spur important investments and jobs in tribal communities that can be leveraged to address some of the impacts from climate change that threaten tribal lands, waters and ways of life" ("Moapa Paiute to Host" 2014).

In August 2015, for $4.3 million, the Moapa settled a lawsuit against NV Energy filed in 2013 with the Sierra Club, alleging that the Reid Gardner Power Generation Station had harmed their health and water supply, thus violating the Resource Conservation and Recovery Act and the Clean Water Act. "For years the band has suffered the consequences of breathing dangerous dirty air from the Reid-Gardner coal plant and this settlement is a step forward," said Senator Harry Reid (D–NV). "While the settlement will provide relief and help make the tribe's home healthier and safer, no amount of money can pay for the sickness caused by a half-century of pollution from the coal plant. The Moapa Band of Paiutes and all Nevadans deserve a clean, healthy environment to raise their families in and pass on to their children" ("$4.3 Million" 2015).

### *WISCONSIN'S MINING MALADIES*

Early in the nineteenth century, Native peoples on mineral-rich lands in Wisconsin, Michigan, and Minnesota were forced off their homelands. Four treaties signed during the 1850s removed 19 million acres of land from Native control as well as 100 billion board feet of timber, 150 billion tons of iron ore, and 13.5 billion pounds of copper. Mining on these lands provided early profits for U.S. Steel under the control of J. P. Morgan. For half a century, beginning in the 1890s, iron mines in the Lake Superior area shipped more than 4 billion tons of iron ore to steel mills in the United States (Gedicks 1993). After World War II, the focus of mining moved to other countries (including Canada, Australia, Venezuela, and Brazil), as the iron-ore mining region near Lake Superior sank into a deep depression.

During the 1970s, miners again focused their attention on this area as the Phelps Dodge Corporation sought a copper-extraction concession from the Lac du Flambeau Chippewa. The company expected speedy approval by the tribal council, but its members, knowing the history of the area, were skeptical. The council hired Charles J. Lipton, a lawyer and expert in minerals-contract negotiation, who told them that "no company that I have seen in recent times would dare to suggest such terms now to a developing nation overseas" (Gedicks 1993). The contract required the tribe (if it was to be a partner) to pay the company $250,000 (worth about $1 million in 2014) to explore its lands for copper; Lipton advised the tribe to decline the offer.

In 1976, Exxon announced the discovery of one of the world's largest potential zinc and copper deposits near the 200-member Sokaogon Chippewa's 1,900-acre reservation near Mole Lake in northeastern Wisconsin. The tiny Indian band soon found itself facing off with the world's largest energy corporation. The tribe had an annual budget of $1,200; Exxon's energy reserves exceeded $1.3 trillion (Gedicks 1993). Deciding that such a mine would ruin their homeland and way of life, the Sokaogon Chippewa decided to resist. The tribe decided that Exxon would be legally vulnerable as it sought permits in areas that threatened their wild-rice harvest. They allied with non-Indians across northern Wisconsin in the Wisconsin Resources Protection Council. After a decade of struggle, the smallest, poorest Indian tribe in Wisconsin leveraged this support into an environmental movement that forced Exxon to withdraw.

By mid-1992, however, Exxon returned, as the prices of both copper and zinc had risen substantially. The estimated value of the deposit had risen from $4.6 billion in 1984 to $7.6 billion in 1992 (Gedicks 1993). The environmental movement also had been weakened by widespread Republican victories in Wisconsin, including the election of Governor Tommy Thompson. Even so, on December 14, 1992, Phelps Dodge withdrew from an anticipated partnership with Exxon in the mining venture.

Also in the mid-1970s, the Kennecott Copper Corporation (a subsidiary of Rio Tinto Zinc, the largest mining company in the world) announced plans to open a small open-pit copper mine in Grant, one mile southwest of Ladysmith, Wisconsin. The Lac Courte Oreilles Chippewa reservation was adjacent to the proposed mine site. Neither Kennecott nor the Wisconsin Department of Natural Resources consulted with this tribe or others, contending that their treaty rights were irrelevant despite contrary opinions in several legal proceedings. The state approved the mine in March 2003.

## *WISCONSIN'S CHIPPEWAS: SULFIDE MINING VERSUS TREATY RIGHTS*

Native Americans in northern Wisconsin have united with neighboring non-Indians to fight a proposal to dig what could become the world's largest iron-ore mine near an area in which native peoples exercise their treaty rights to hunt, fish, and gather wild rice. The political movement that has grown out of resistance to mining in this area, which calls itself the Watershed Alliance to End Environmental Racism, went head-to-head with energy-industry lobbyists in the Wisconsin statehouse over a proposed mining moratorium. Opposition to mining intensified as Wisconsin residents discovered what the mining wastes could do to the Earth from which they wrest their lives. Metallic sulfides, combined with air and water, can create sulfuric acid as well as residues of several toxic metals, including mercury, lead, and arsenic.

Dan Kaufman (2004), writing in the *New York Times*, sketched what is at stake in far northern Wisconsin's Penokee Hills:

> The $1.5 billion mine would initially be close to four miles long, up to a half-mile wide and nearly 1,000 feet deep, but it could be extended as long as 21 miles. In its footprint lie the headwaters of the Bad River, which flows into Lake

> Superior, the largest freshwater lake in the world and by far the cleanest of the Great Lakes. Six miles downstream from the site is the reservation of the Bad River Band of Lake Superior Chippewa, whose livelihood is threatened by the mine.

Plans for the mine call for the dumping of waste into a 350-acre waste pond that would reach, in some locations, to within 15 feet of the water table, raising fears that sulfuric acid and heavy metals would leach into local water supplies. Not only are these waters used for bathing, drinking, and cooking, but they also sustain the local Chippewa with fish as well as wild rice, a cash crop as well as a sacred part of Sokaogon Chippewa religious life. Native Americans in the area realized that if the mine is constructed, their drinking water and their harvest could become inedibly toxic.

The conflict over sulfide mining in Wisconsin began in 1975, when Exxon Minerals Company located one of the largest zinc–copper sulfide deposits in North America adjacent to the Mole Lake Sokaogon Chippewa reservation near Crandon, Wisconsin. Company plans anticipated the extraction of roughly 50 million tons of sulfide ore over the life of the mine, about 30 years. This volume of waste could fill Egypt's Great Pyramid about 10 times.

Once word of the mine's possible effects spread, sportsmen's groups, many of which had opposed local Native American fishing rights a few years earlier, allied with Native Americans because their access to healthy wildlife was threatened. The Chippewas also have been joined by several non-Indian environmental groups, including the Green Party. The environmental alliance against sulfide mining in the area began in earnest during the 1980s, when Kennecott Copper, a subsidiary of Rio Tinto Zinc, developed plans for an open-pit copper sulfide mine near Ladysmith. In 1993, a small mine was started on this site while local opponents feared that a larger mine would follow. The proposed mine site lies on land guaranteed to the Sokaogon Chippewa by treaty in 1855.

Mine supporters have maintained that new technologies will mitigate the spread of sulfide by-products from the mine wastes, but few people in the area bought the company's argument. Mine opponents worked with State Representative Spencer Black (D–Madison) to fashion a bill that would prohibit the opening of a sulfide mine in Wisconsin until a similar mine in another state had been closed for 10 years with no contamination of surrounding groundwater. The bill passed the state senate, 29–3, in March 1997; in late 1997, the bill was reported out of the state assembly's Environment Committee by a 6–4 vote. "There are enough votes to pass it; [but] the question is rather the bill's opponents will be able to attach weakening amendments during the assembly floor debate," said Al Gedicks, a professor of sociology at the University of Wisconsin, La Crosse (Johansen 1998, 6). The bill never passed, and planning for the huge open-pit mine continued.

## *WISCONSIN: PENOKEE HARVEST CAMP*

The Lac Courtes Oreilles Band of Ojibwe and several supporters established the Penokee Harvest Camp in March 2013 to oppose a proposed open-pit iron-ore mine by Gogebic Taconite (GTAC) adjacent to the Bad River reservation six miles from the proposed mine site. Occupants of the camp and their supporters maintained that the proposed open-pit mine would pollute the Bad River watershed, which drains into Lake Superior, damaging wild rice, fish, and wildlife. The mine, as proposed, eventually could reach 22 miles in length.

The iron-ore mining battle became a visceral, racist conflict that spilled into social media. A mainly white group, Wisconsinites for Safe Mining, became a font of racist cant, as described by Mary Annette Pember for the Indian Country Today Media Network:

> A [Facebook] post that has since been removed includes a historical photograph depicting a black and white portrait in the Edward Curtis style of a stoic-faced Native man dressed in regalia. The author has placed the following caption on the top and bottom of the image, "Tries to Explain How Natives Are a Proud People, Too Drunk to Form Words." In a post responding to disagreement with the mine, Wisconsinites [for Safe Mining] responds, "We are still going to mine the hell out of Wisconsin, that is not going to change. If you protesters get too close to the hole, we'll push you in! Today's posts are brought to you by Dirty Harry. Because a .44 Magnum would blow your head clean off. Do you punks feel lucky?" (Pember 2013c)

Mike Wiggins Jr., chairman of the 2,000-member Bad River tribe, looked at the conflict as a "clash of cultures." However, he said, "you have to see life from the eyes of a sturgeon. . . . Rivers and streams are the lifeblood of the Earth." While many members of the tribe believe that the mine will foul their river and the Earth with mercury, asbestos, sulfuric acid, and arsenic, Seitz, speaking for the mining company, "envisioned reclaimed land becoming an ATV park or a long shooting range, [with] people and fish swimming in the mine-made lake" (Pogash 2013, 4-A).

The Iron County Forestry Committee met on July 23, 2013, and voted during a closed session to declare the opponents' camp illegal because its members had failed to obtain a permit. The committee authorized legal action to evict the campers. The Penokee Harvest Camp residents said that they were legally exercising their treaty rights to fish, hunt, and gather. The county asserted that camping for more than two weeks without an "extended permit" was not among their treaty rights, while camp members believed that it was implied. (It's very difficult to hunt and gather without camping.) Workers for GTAC were conducting exploratory drilling in the area and may have been seeking the campers' eviction. The Indian Country Today Media Network reported on July 20, 2013, that "LCO [Lac Courtes Oreilles] tribal chairman Mic Isham noted that the county permitting process is essentially a non-issue since Ojibwe are allowed by treaty to hunt, fish and gather in that area, which is part of ceded territory. He points out that the tribe had sought Iron County permission only to ensure that non-Native visitors would also be allowed to camp there." The Indian Country Today Media

Network also reported that "Harvest Camp spokesperson Paul DeMain indicated that if necessary all tents at the camp could be taken down every 14 days for at least an hour and set back up in order to comply with Iron County requirements" (Pember 2013e).

By mid-2013, Gogebic Taconite was maintaining eight test-drilling sites in the Penokee Hills. Rob Ganson hiked in the area with a camera. On July 6, walking slowly because of his emphysema, Ganson, 55, said he was surprised when he and his four friends approached the drill site and encountered a man dressed head to foot in camouflage and carrying an assault rifle. Ganson was even more surprised when a second man suddenly appeared behind them in a similar outfit, also carrying an assault rifle. "I was kind of concerned for my wife," he says. "I didn't much like her walking into the midst of men carrying machine guns." As Ganson and his friends approached the site, a man met them at the entrance, clearly guarding the drilling area. He simply stood there facing them and holding an automatic rifle.

The armed guards were the company's response to an incident in June 2013, when protesters from Earth First! wearing black head coverings had invaded the mine site, cursing at workers, slashing tires, and damaging GTAC equipment worth about $2,000. The company called the attack "eco-terrorism" (Pember 2013d). After that, in July, GTAC brought in armed guards from Arizona's Bulletproof Securities. Bob Seitz, spokesperson for GTAC, confirmed that the company had hired security for the drilling sites and that the men have been patrolling the hills for a while. "These people aren't supposed to be seen," he said. "This is probably the first time that visitors from the Harvest Camp have spotted them." In a July 8 story in the *Wisconsin State Journal*, State Senator Bob Jauch (D–Poplar) said that the guards are from Bulletproof Securities, an Arizona company that boasts a "no compromises security force" (Pember 2013a).

On March 11, 2013, Republican Governor Scott Walker signed into law a "streamlined" permitting process that eliminated several environmental regulations to speed the development of new mines. The law reduced the number of public hearings required for approval of the open-pit mine proposal and required speedy state responses to environmental impact statements. This law also allowed miners to dump millions of tons of waste into the Bad River watershed, stating that "a significant adverse impact on wetlands is assumed to be necessary" (Pogash 2013, 4-A). Democrats called the bill "a giveaway to an out-of-state corporation of Wisconsin's lands and waters" that enables a slow poisoning of his people with pollutants that would be tantamount to genocide. To Walker, the bill was a vehicle to "create sustainable jobs for generations" (Hughes 2013a). GTAC's executives and other supporters of the proposed mine have given about $15 million to Governor Walker, as well as Wisconsin Republican legislators, 600 times their donations to the mine's opponents (Kaufman 2014).

On March 15, 2013, Wiggins said that a legal defense fund had been established to oppose permits for the iron-ore open-pit mine that he said could eventually become the largest of its type on Earth. The Lake Superior Chippewa also were defending a 16,000-acre wetland in the Kakagon and Bad River sloughs in an alliance with several environmental groups. A Wisconsin Wetlands Association study conducted with the National Oceanic and Atmospheric Administration said that "the exceptional health of these wetlands is owed to the stewardship and protection provided by the Bad River Band. Despite this protection, the sloughs are still vulnerable to external threats, most notably mining in the Penokee Hills at the top of the watershed, which could irreparably alter hydrology, water quality, and wildlife habitat" (Hughes 2013a).

Having lobbied its mine proposal for several years, pouring several million dollars into its campaign in exploratory expenses and political contributions, in February 2015, Gogebic Taconite suddenly closed its office in Hurley, Wisconsin, and pulled out of the area, leaving behind a galvanized opposition that had united Native Americans and whites whom heretofore had agreed on little else; an Ashland County Board that had tempered its initial support with regulations of prospective noise, damage to roads, dust, and pollution; and Wisconsin's governor, Scott Walker, with a $700,000 GTAC donation in his political coffers.

No one was saying that anything was final. The company was reacting in part to a petition by the Wisconsin Federation of Tribes requesting that the EPA use the Clean Water Act (as it had to stop the Pebble Mine in Alaska) against GTAC's proposal. "It's unfortunate that the federal requirements for mitigating wetlands make it cost-prohibitive for Gogebic to move forward at this time. We remain committed to working with companies interested in creating quality, family-supporting jobs in Wisconsin," Walker said. "We will continue to investigate the possibility of pursuing a permit to mine the Upson site (in Iron County) but cannot justify maintaining an office in Hurley without a prospect of immediate action," Gogebic added. Opponents of the mine were not declaring victory. "Mines are like vampires; they're very hard to kill. The mineral deposit [that attracted GTAC] isn't going away," said Rich Eggleston, a member of Citizens Concerned, about the proposed Penokee Mine (Pember 2015).

## *TRIBES OPPOSE MINING IN THE TONTO NATIONAL FOREST, ARIZONA*

For several years, Arizona tribes have been resisting mining in the Tonto National Forest that would replace land they hold sacred with a two-mile-wide, 1,000-foot-deep strip mine. In 2013, U.S. Representatives Paul Gosar (R–AZ) and Ann Kirkpatrick (D–AZ) reintroduced the Southeast Arizona Land Exchange and Conservation Act of 2013, which had failed in the U.S. Senate a year earlier, to enable construction of the Superior Mine. The latest attempt was their eighth since first introducing this bill in 2005. In 2014, the same land exchange was introduced again by Republicans in the U.S. House of Representatives as part of the National Defense Authorization Act, "a must-pass bill that sets the nation's defense policy" (Wilson 2014).

The bill would order the transfer of more than 2,400 acres of Oak Flat Campground and nearby Tonto National Forest land into private ownership by

Resolution Copper Mining, a subsidiary of Rio Tinto PLC of the United Kingdom and BHP Billiton Ltd of Australia. About 5,500 acres of nonfederal land would be added to the forest in exchange, according to the Indian Country Today Media Network which reported: "Tanayia White, assistant to Terry Rambler, chairman of the San Carlos Apache Tribe, said the proposed exchanged lands are no substitute for the Oak Flat area, held sacred by her tribe. And the biodiversity at Oak Flat is unmatched, she said—even when it's held against the riparian acreage, so precious in the West, that would be traded for it." Representative Gosar countered that the exchange "will create over 3,700 high-paying jobs, increase U.S. energy and mineral independence, and preserve some of rural Arizona's most beautiful natural lands" (Minard 2013).

The San Carlos Apache have allied with a large majority of Native tribes and nations in Arizona and Nevada as well as the National Congress of American Indians, the Great Plains Tribal Chairman's Association, the Affiliated Tribes of Northwest Indians, the All Indian Pueblo Council of New Mexico, and the United South and Eastern Tribes. Environmental groups (the Access Fund, the Arizona Mining Reform Coalition, the Tucson Audubon Society, the Friends of Ironwood Forest and Earthworks, and citizen groups, including the Concerned Citizens and Retired Miners Coalition and the Arizona Mining Reform Coalition) have joined the protest to stymie the land transfer for several years.

"RCM [Resolution Copper Mining] has indicated it will use the block-cave mining technique to extract the copper from Arizona public lands, a process that will destroy huge swaths of land in the Tonto National Forest and consume more than 40,000 acre feet of water yearly," wrote the coalition. "In addition to the massive water withdrawal, the process will release toxins through the mining process that can contaminate and further deplete Arizona's precious and limited water supply" (Minard 2013).

Many opponents look at the proposed land transfer as the worst kind of government and corporate collusion to rape the land. Don Steuter, conservation chair for the Sierra Club's Grand Canyon Chapter, asserted that the bill "allows Rio Tinto to privatize public, sacred lands, including Oak Flat, which are of incalculable value to Native Americans, birders, rock climbers, and endangered species. And it does this by sidestepping a cornerstone of our environmental laws—the National Environmental Policy Act." Roy Chavez, of the Concerned Citizens and Retired Miners Coalition, said, "If passed, this bill may leave Arizona with nothing but a massive hole in the ground and a huge cleanup bill costing the American taxpayers billions of dollars" (Minard 2013).

By 2015, the miners were still after the land, as 300 Apaches and supporters marched 44 miles from San Carlos tribal headquarters to a camp on their sacred land that they plan to occupy to retain the area. "The deal is an impressive new low in congressional corruption, unworthy of our country's ideals, no matter what side of the aisle you're on," commented Lydia Millet (2015, A-27) in the *New York Times*. She noted that the latest version of the land exchange was written so that it would take effect no matter what a required environmental assessment said. If this was a Christian holy site, members of Congress might think twice before allowing it to be turned into a strip mine, wrote Millet. "But this is Indian religion" (Millet 2015, A-23).

## *MINING A HISTORICAL SITE ON NORTH DAKOTA'S KILLDEER MOUNTAIN*

Taĥċa Wakutėpi (Killdeer) has long been a place where young Lakota and Dakota men seek visions in the *Haŋblėċiyė* ("Crying for a Vision") ceremony. In 2013, oil companies proposed drilling for oil here, part of the Bakken oil field boom, one of the largest fossil-fuel excavations in U.S. history. Native peoples quickly rallied against the proposal because Killdeer is not only a traditional sacred site but also a mourning ground. Into this natural fastness, as in so many areas of the West, comes a proposal to turn Mother Earth to mother lode, with drilling by the Hess Corporation for oil and natural gas.

On June 28 and 29, 1864, on the Killdeer Battlefield (now a national historical monument) near Killdeer, North Dakota, Brigadier General Alfred Sully led about 4,000 soldiers in the last days of the Great Sioux Uprising. Sully and his army were far afield of the war's original battle sites in Minnesota that had ended with the largest mass hanging in U.S. history: 38 people in Mankato, Minnesota, in 1863. Sully's forces attacked a summer gathering of peoples from the surrounding area who were socializing, trading, and taking part in ceremonies, killing about 150 Dakota and Lakota men, women, and children. In his version of total war, General Sully directed his troops to destroy as many as 1,800 lodges; 200 tons of buffalo meat and dried berries, clothes, and household utensils; tipi poles; travois; and piles of tanned hides and slaughtered horses and perhaps 3,000 dogs. Sitting Bull and Gall, who fought at the Battle of Little Bighorn in 1876, were among the survivors (Lee 2013).

The Indian Country Today Media Network described Killdeer Mountain as

> hardly a mountain, but it is a beautiful and majestic plateau as it rises gently above the steppe of the Northern Great Plains. In the summer, native plants and flowers dot the hillside and grow in the cracks of shattered sandstone. Short and middle indigenous grasses sway in a wind that has been present since creation. The song of coyotes hauntingly fills the air on a gentle midsummer's eve. The trees, a mix of ash and cottonwood, grow in clusters, but it's the cottonwood trees which sway and shush the world. Crickets take up their hum in the twilight where the cicadas left off theirs in sunlight. Aeries of golden eagles and hawks remind the meadowlarks and rabbits to keep a wary eye on the skies. One golden eagle circles lazily above me and I take it as a good sign, my prayers will be carried, and I pause a moment to remember my grandparents.
>
> At the very top of the plateau is a cave, an entrance into the heart of grandmother earth. Medicine Hole. Since the days of warriors and legend the *Nu'Eta* (Mandan) have called the mountain *Bah-eesh*, the Mountain That Sings. By day, like a great inhalation, the wind rushes into the deep embrace of the

> earth and at night like a long sigh the wind comes out with a whistle, and if one listens carefully, the song of the earth. (Dakotah 2013)

In the summer of 1998, Isaac Dog Eagle performed a "Releasing of the Souls" ceremony on the site of the Killdeer attack by Sully's forces. A year later, Dog Eagle also performed a "Wiping of Tears" ceremony as a healing for people who had lost family members in the attack.

Richard Rothaus of the archaeological consulting company Trefoil Cultural and Environmental has found three sites containing human artifacts and remains on the mountain, two of them major. According to Rothaus, "Oil development is growing so fast out there that no one can keep track of it. People didn't know this was being leased," so they couldn't do the archaeological work earlier. "Hess says it avoided the battlefield, but is 2.5 miles away from the historical marker for the battlefield, not three miles away from the battlefield" (Lee 2013).

Gerard Baker, an enrolled member of the Three Affiliated Tribes and a former National Park Service superintendent, was 59 years of age in 2013. As a child, he learned the ceremonial importance of Taĥċa Wakutėpi from his father, who learned it from tribal elders. "He told us the stories of Singing Butte, where Earth-naming ceremonies once took place. Many of the ceremonies are lost because of time, but they are still extremely important. Medicine Hole is associated with lost ceremonies. Many were lost during the smallpox epidemic of 1837" (Lee 2013). Baker explained that unless a ceremony's owner sells or gives away the ceremony before he dies, it can no longer be performed. So many Indians died so quickly during the smallpox epidemic that they did not have time to ensure the survival of their ceremonies.

The spirits still live on Singing Butte, however. "The spirits live in different areas throughout the Dakotas in various buttes from Canada to the South Dakota line. The Hidatsa consider that their ancestral territory," said Baker. Burials, fasting, trapping to get eagle feathers, deer hunting, and dressing also took place there. A fundamental problem—and one of the challenges in opposing oil drilling on the mountain—said Baker, is that "not enough people know about the ceremonies. Even though people know the site is sacred, not so many know about the ceremonies." He has a very pragmatic approach to dealing with Hess's current oil drilling proposals. "I wish we could say 'No drilling,' but that's not going to happen. They're going to get that oil one way or the other. . . . We could hold up protest signs, but I think education would work better," said Goodhouse. "I feel the issue is people don't care because they don't know" about Killdeer Mountain's cultural or historical significance (Lee 2013).

### *MONTANA'S GROS VENTRE AND ASSINIBOINE, GOLD MINING, AND CYANIDE POISONING*

The Little Rocky Mountains of Montana, which long have been regarded as sacred by the Assiniboine and Gros Ventre, are now laced by the effluent of open-pit gold mines that have produced toxic acid mine drainage. Andrew

Schneider of the *Seattle Post-Intelligencer* described Gus Helgeson, the president of Island Mountain Protectors, a Native American environmental and cultural organization, standing atop Spirit Mountain as he scanned "the gashes, pits and piles of rock that once was his tribe's most sacred land" (Schneider 2001). "The strong man weeps," Schneider wrote.

In 1855, the Assiniboine and Gros Ventre were moved to the Fort Belknap reservation, which was named for a U.S. secretary of war. The Assiniboine and Gros Ventre gave up 40,000 acres of land in exchange for a government promise to feed, clothe, and care for them. At the time, federal Indian agents said nothing about the gold that was buried in Spirit Mountain, but they made it clear that the tribes could either agree to their terms or starve. Spirit Mountain is part of the Little Rockies, an island of mountains in the nearly flat prairie. To Native Americans, the mountains were valued for their deer, bighorn sheep, herbs, natural medicines, and pure water. Gold mining has destroyed all of that.

In 1884, Pike Landusky and Pete Zortman discovered gold on the reservation. Facing starvation, in 1895 the Assiniboine and Gros Ventre signed an agreement negotiated by General George Bird Grinnell, selling portions of their gold-laced land to the federal government for $9 an acre worth of livestock and other goods, as miners besieged towns that had been named after the gold's discoverers. Under the terms of the General Mining Law of 1872, the government sold the land to individuals and private companies for $10 an acre.

"The first time the mining company let me up here, let me see what they had done to our land, the pain surrounded me. It was like watching our ancestors die, raped of their honor," said Helgeson. "They destroyed this place, took their gold off in armored trucks and left us a wounded mountain spewing poison on the people the mountain was stolen from," he said (Schneider 2001).

Over several decades, according to Schneider's account, "scores of shafts were driven into the Little Rockies, and an estimated $1 billion in gold and silver was taken out of the ground—more than $300 million by the last owner of the mine, Pegasus Gold Corp. of Canada" (Schneider 2001). Underground mining continued until the 1950s, after which open-pit strip mining was initiated. In 1979, as gold prices rose rapidly, the Pegasus Gold Corporation and a subsidiary, Zortman Mining Inc., built mines that extracted gold from heaps of low-grade ore with cyanide solutions.

Pegasus Gold, which owned several mines in the Little Rocky Mountains, went bankrupt when gold prices fell sharply after 1980, "leaving the state of Montana with a $100 million cleanup liability and the tribes with the prospect of perpetually polluted water" (Huff 2000). Cyanide-assisted gold mining continued until 1990, during which time the mine was expanded nine times without any substantial environmental review, despite cyanide spills into the water table used by the Indians (Abel 1997).

By 1990, the Assiniboine and Gros Ventre began to challenge the environmental side effects of cyanide-heap gold mining, forming a native environmental advocacy group, Red Thunder, which joined with non-Indian environmental groups to resist federal permits for the Zortman–Landusky Mine's next requested expansion.

The group's appeal was denied. By December 1992, Pegasus had applied for another expansion of the mine, as Red Thunder joined with another Native environmental group, Island Mountain Protectors. Both prepared plans to challenge the expansion under the federal Clean Water Act, maintaining that the cyanide-leach method used in the mine was poisoning the reservation's water supply.

In July 1993, a heavy thunderstorm brought a flood of acidified mine wastewater into the town of Zortman, after which the U.S. Bureau of Land Management required the mine's owners to develop a new reclamation plan. At about the same time, an EPA study found that the mine had been "leaking acids, cyanide, arsenic and lead from each of its seven drainages" (Abel 1997). The state of Montana soon joined the EPA in a suit based on the Clean Water Act, which was settled out of court in July 1996 with Pegasus and Zortman Mining pledging to pay $4.7 million in fines to the tribes, the federal government, and the state. The mine's owners also pledged to follow a detailed pollution-control plan in the future. Shortly thereafter, a request to triple the mine's size (from 400 to 1,192 acres) was approved by the Montana Department of Environmental Quality and the Bureau of Land Management (Abel 1997). In January 1997, the Fort Belknap Community Council, the National Wildlife Federation, and the Montana Environmental Information Center sued the Montana Department of Environmental Quality, alleging that the agency's decision to allow an expanded mine violated state law.

In September 1997, federal and state environmental agencies fined Pegasus and Zortman Mining $25,300 for violating the settlement by polluting a stream in the Little Rockies the previous summer. John Pearson, director of investor relations for Pegasus, asserted that discharges were the result of "acts of God" during "extraordinarily heavy rains" (Abel 1997). By late 1997, with gold prices (and its share price) declining rapidly, Pegasus warned that its mine would close by January 1, 1998, if the expansion plan were not accepted. The state of Montana and local Native environmental activists wondered whether Pegasus would survive long enough as a corporate entity to complete the promised reclamation of existing mines. In January 1998, Pegasus filed for Chapter 11 bankruptcy protection.

In the meantime, Pegasus left behind open pits that were described by Schneider (2001):

> Pegasus dug pits the size of football fields and lined them with plastic or clay. Crushed ore was dumped in mounds as high as 15 feet and soaked with a mist of cyanide. It was the largest cyanide heap–leach operation in the world. . . . The heavily contaminated water trickled and flowed through fissures in the mountain, into the surface streams and underground aquifers that supply drinking water for 1,000 people who live in and around Lodge Pole and Hays, reservation towns north of the mountains.

Streams flowing off the mountain smell of rotten eggs (the chemical signature of sulfide), cloudy and lifeless. "This is death," said John Allen, a tribal spiritual leader, as he filled his hands with putrid muck. "The mines take millions in gold from our land and leave us poisoned water. The miners and the government experts have argued for years about whether the water is bad. All they have to do is look, but they choose not to see" (Schneider 2001). Allen, who was 46 years of age in 2001, has thyroid problems, as do three of his siblings. His father has lymphatic cancer. Doctors who specialize in environmental medicine have told the Assiniboine and Gros Ventre that the diseases they suffer stem from contaminated water. Environmental advocates among the two tribes also report a high rate of stillbirths.

### *THE HUICHOL (WIXÁRITARI), A SACRED SITE, AND SILVER MINING*

The Huichol (Wixáritari) reside near silver deposits in the Cerro Quemado, a sacred site adjacent to their reserve in Mexico: "the mountain called Cerro Quemado . . . birthplace of the sun . . . rises up out of the land shrouded by green shrubs and cactuses and towers over the natural marvel that is this landscape. The Huicholes have been making an annual pilgrimage to the Cerro Quemado from their native lands some 250 miles to the west for as long as anyone can remember" (Zhorov 2012b). Silver also has been mined there since the eighteenth century. The mines closed in 1991, and Real de Catorce, one of the larger towns in the area, began making money as a tourist destination. The Huichol have (not always willingly) become the object of a New Age invasion by peyote-ingesting Norte Americanos. Tourism declined with rising narcoviolence in Mexico, however, and Real de Catorce's population declined to less than a thousand people.

By 2012, with silver prices rising, the miners returned. Minera Real Bonanza, a subsidiary of the Canadian First Majestic Silver Corporation, began a project it calls La Luz ("The Light"), an expanded version of the old mining system. Its object: $3 billion to be extracted from ore. Locals, while desiring jobs that mining may bring, also fear that a large-scale mine will drive away tourists. The real object of the controversy is the nature of mining, which has changed. Older, labor-intensive subsurface mining may be replaced by an open pit, as labor costs decline, along with the ratio of silver ore to "overburden," or waste rock. This is true throughout the mining industry worldwide.

Royalties often are not paid to local people in Mexico, and open-pit mining may ruin the land and water. Once the life of the mine ends (about 15 years), local people may be living on a wasteland. Minera Real Bonanza, however, promises 500 direct jobs that would generate 1,500 more indirectly after work begins. The company also says it will mine discreetly, that "mine entrances and tailings will be tucked into the landscape and invisible to tourists. . . . The company has proposed to build a water-treatment plant—Real's first—to treat 'black water' instead of expulsing it directly into streams, as is done now. The company is preparing a closing plan, which will involve total remediation of the site" (Zhorov, 2012b). The company also proposed an exhibition mine for tourists and a mining museum.

Some of the company's concessions are result of activism by the Zapatista uprising that made indigenous rights a political issue in Mexico. In 2008, the Hauxa Manaka Pact for the Preservation and Development of the Huichol Culture was signed to protect the Huicholes' sacred places and culture generally. The Defense Front of Wirikuta held educational forums and fund-raising concerts, drawing as many as 50,000 people to stress the sacred nature of Cerro Quemado and other aspects of their peyote culture. The mining company says that Cerro Quemado will not be strip-mined.

"In the company's offices hangs a piece of Huichol art depicting a slice of Huichol mythology, all psychedelic colors and dreamscapes," reported Irina Zhorov of the Indian Country Today Media Network. However, not everyone trusts the company's promises. María Teresa Sánchez Salazar, who teaches in the geography department of the National Autonomous University of Mexico, said that throughout the country, promises and laws aren't the problem. "The issue is for them to actually abide by the legislation," she says. "There isn't supervision strict enough to make them fulfill [their obligations]" (Zhovov 2012b).

## *THE ENVIRONMENTAL ECONOMY OF KITTY-LITTER STRIP MINING*

Representatives of the Reno-Sparks Indian Colony paid a surprise visit to the 2001 annual meeting of the Oil-Dri Corporation shareholders' meeting in Chicago to stop a proposed kitty-litter strip mine at the edge of their 550-person community in Nevada. The residents of the community, in Hungry Valley in central Nevada, wanted to prevent Oil-Dri from opening two strip mines and a processing plant within a mile of their community. They fear heavy traffic near schools, pollution of groundwater, and dust-laced air pollution from the mines and processing plant. They also presented evidence, supported by an array of photographs, that reclamation will never heal the Earth in the dry environment.

Ben Felix, a Hungry Valley citizen working with the colony, said in Chicago that "six of us came from Reno, Nev., to attend the shareholders' meeting of the Oil-Dri Corp. They are proposing to put a strip mine and manufacturing plant in an already existing residential area. This directly affects the Reno-Sparks Indian Colony. They are trying their utmost best to get this processing plant that will run 24 hours a day, seven days a week for 365 days out of the year" (Pierpoint 2001, A-1).

"They are claiming that they will be able to reclaim the existing land back to its pristine nature," said Felix, "but it really doesn't address the issues. Arsenic, mercury, selenium and some other minerals will be leached into our groundwater. The air, dust and pollution that will be caused in the valley will mean breathing very, very fine particulates, finer than chalk dust off a chalkboard, and this is the air our children are going to have to breath for the next hundred years" (Pierpoint 2001, A-1).

Oil-Dri asserts that the Reno–Sparks Mine is the only commercially viable mine for cat litter in the United States. "If this [the proposed mine] is exhausted, does that mean the end of kitty litter as we know it?" asked Felix. "Are our children going to have to suffer severe asthma and other health problems just to satisfy the pocketbook of one corporation?" (Pierpoint 2001, A-1).

Oil-Dri has received reclamation and air permits for the projects during the two and a half years that the Indians have resisted it. The company planned to spend $10 million on the mines and processing plant, which it says will offer as many as 100 jobs. Washoe County planning commissioners, however, voted on December 19, 2001, to deny a permit allowing Oil-Dri to open a kitty-litter processing plant in Hungry Valley.

## *THE CHEYENNE RIVER SIOUX AND THE HOMESTAKE MINE'S TOXIC LEGACY*

The Cheyenne River Sioux, with legal support from the U.S. federal government, have successfully sued the Homestake Mining Company for polluting waterways in South Dakota with lead-laced mining waste. Homestake was sued during 1997 for dumping 100 million tons of mine tailings from the Lead (South Dakota) Mine, which had been contaminated with cyanide and arsenic into Whitewood Creek, the Belle Fourche River, the Cheyenne River, and the Missouri River during the previous century. The company asserted that the dumping stopped in 1977.

According to a report in *Drillbits and Tailings*, the Lead Mine is located northeast of the Cheyenne River Sioux reservation. It was the first large-scale gold mine in the United States, containing deposits discovered by George Armstrong Custer's incursion into the Black Hills in 1874. Over a century, this mine probably produced the single largest quantity of gold of any mine in the Western Hemisphere ("Cheyenne River" 1999).

Homestake was ordered to pay a total of $5 million, most of which was directed toward restoring the Whitewood, Belle Fourche, and Cheyenne rivers. One-third of the $4 million was reserved for restoration projects on the Cheyenne River Sioux's reservation; the rest was be placed in the South Dakota Game, Fish, and Parks Fund to be used jointly for restoration by the United States and South Dakota. "This is the first time Homestake has ever had to pay anything for any of the damage they've caused," said Steven Emery, Cheyenne River Sioux tribal attorney ("Cheyenne River" 1999).

## *NAVAJO COAL: ASSET OR DEBIT?*

Navajo environmental activists have asserted that coal production and power combustion there have long been "causing respiratory illnesses, depleting local aquifers and contaminating drinking and agricultural water with chemicals. . . . The coal mining is like a knife cutting into the liver of our mother land," Wahleah Johns, the solar project coordinator for the Black Mesa Water Coalition told New America Media ("Navajo Activists" 2012). "The liver is what keeps the whole body healthy. The coal mining is why our people are getting sick." Ruben Hernandez (2012) wrote that Navajo activists are utilizing "a cultural imperative to protect the environment and serve as stewards of the land, harnessing their natural resources, like the area's scorching sun and thrashing winds" ("Navajo Activists" 2012).

The Navajos also have been forming alliances with Latinos in Arizona and New Mexico. Census figures indicate that Phoenix and Tucson were more than 40 percent Latino by 2012, and some have been organizing politically on issues that combine economic parity, immigration reform, and sustainable energy. By 2012, the Navajo Nation was erecting 48 wind turbines in its Big Boquillas Wind Project about 80 miles west of Flagstaff, Arizona. Wind sites also are being scouted in the Four Corners area (at the conjunction of Arizona, New Mexico, Colorado, and Utah), heretofore known for a coal-generated power plant so large that NASA satellites have located its plume from Earth orbit.

Even so, coal mining and power generation remain among the largest contributors to the Navajo Nation's budget. As the Indian Country Today Media Network reported,

> The Navajo Generating Station (NGS), a coal-fired plant in Page, Arizona, is operated by a group of six regional utilities including the Salt River Project (SRP), one of Arizona's largest utilities. From 1987 to 2010, the NGS and the nearby Kayenta Mine, which provides roughly eight million tons of coal per year to the NGS, contributed nearly $1.3 billion to the Navajo and Hopi economies, according to a National Renewable Energy Laboratory (NREL) report. The 120-page federal report released in January 2012 reveals that wages and benefits paid to Native American employees at the NGS amounted to approximately $52 million per year for those 23 years. The average hourly wage paid at the power plant is $35 per hour—twice the average wage paid in the county, according to Bureau of Labor statistics. The NREL report also notes that 83 percent of the NGS workforce, about 450 employees, is Navajo. With the unemployment rate exceeding 50 percent on the reservation, these jobs are extremely valuable, SRP and some tribal government officials argue.

Navajos and non-Native environmental activists have brought pressure to bear to reduce emissions from the Navajo Generating Station, but as of late 2013, it was still polluting the area and compromising people's health. The National Park Service argued that pollution over the entire region should be reduced. "For decades this plant has emitted massive amounts of preventable pollution into the skies above our national parks like the Grand Canyon, Petrified Forest and Mesa Verde, as well as into the lungs of hundreds of thousands of local residents and visitors to these magnificent places," said the Arizona program manager for the National Parks Conservation Association, Kevin Dahl, on November 11. "The pollution from this plant must be substantially reduced as soon as possible, for the sake of our lungs and our parks" ("National Parks" 2013).

An agreement reached between the Navajo Nation, the Department of the Interior, and several other groups in mid-2013 allowed the plant to continue operation on an extended Navajo lease but with much-reduced pollution levels. The plant is one of the largest sources of nitrogen oxides in the United States as well as other sources of pollution.

### *A COAL MINE PURCHASE BY THE NAVAJO*

Even as the Navajo Nation was banning uranium mining and milling as well as debating the health effects of coal mining and electricity generation, in December 2013 it was buying an open-pit coal mine that supplied the Four Corners Power Plant. Supporters of the purchase said it gave Navajos some control over their reservation's resources. Opponents said the Navajos were buying into another toxic legacy. Some Navajos called on the Department of the Interior to reverse the $85 million purchase south of Farmington, New Mexico.

"Coal is a dirty business and it's already impacting negatively on our health," said Duane "Chili" Yazzie, president of the Navajos' Shiprock Chapter. "The coal market itself is declining, so the overall trend is away from fossil fuel. The Environmental Protection Agency seems like they're going to get more and more stringent, so why do we want to buy into a dying industry?" (Landry 2014).

"Rather than sitting on the sidelines, we now have a say in the energy industry in terms of how that reserve of coal is being used," said LoRenzo Bates, chairman of the Navajo Nation Council's Budget and Finance Committee. "This say goes beyond the coal industry and allows us a voice in alternative forms of energy. Coal can have other uses, and this makes us a player in the industry." Advocates of the mine point out that it employs about 800 people on a reservation where unemployment averages 50 percent (Landry 2014).

### *TURNING BLACK MESA'S WATER TO COAL SLURRY*

Slightly less than half of the electricity consumed in the United States is generated by burning coal. Some of that coal comes from strip mines on American Indian reservations. One major example is the electricity-generating complex on the Navajo Nation at the Four Corners area. "Since 1974," according to one observer, "the Mojave Generating Station and the Navajo Generating Station . . . have been polluting the world's air. The Mojave Generating Station alone uses 18,240 tons of coal per day at full load. Combined, the two plants require 12 million tons of coal a year and are the largest polluters in the country. Astronauts saw the pollution cloud from these coal fired plants from the Moon" (Giuliano 2001). The six Four Corners coal-fired power plants, completed during the 1970s, emit pollution that on some days fills the local atmosphere with higher levels of sulfur and nitrogen oxides than the air of New York City or Los Angeles (Schneider n.d.).

Coal-fired power plants emit more toxic pollution than any other form of electricity production. For every megawatt hour of electricity produced, coal generates 2,071 pounds of carbon dioxide, 13.8 pounds of sulfur oxides, 4.8 pounds of nitrogen oxides, and 3.2 pounds of particulate matter. By comparison, natural gas emits 1,205 pounds of carbon dioxide per megawatt hour, 0.008 pounds of sulfur oxides, 4.3 pounds of nitrogen oxides, and negligible particulate matter (Giuliano 2001).

A significant proportion of the coal that powers these plants is strip-mined from Black Mesa, Arizona, home of the Hopi Indian reservation (which is also home to several thousand Navajo), site of a growing Peabody coal mine that is steadily conveying the area's water resources out of this arid area as coal slurry. On Black Mesa, the Peabody Coal Company uses more than 3 million gallons a day (1.4 billion gallons a year) of once-pristine, potable groundwater that leaves the aquifer as coal slurry.

The aquifers that in 2000 provided 60,000 people on Black Mesa with water were being depleted much faster than nature can replenish them. Computer models run by hydrologist Ron Morgan indicate that by 2050, not even allowing for Peabody's coal-slurry needs, "virtually every spring on the Hopi homelands will be bone dry." By 2050, Morgan estimates that the aquifer will be drawn down at 10 times the rate that it is being recharged. "All the computer models tell us that these depletions are right around the corner," Morgan has been quoted as saying (Vandevelder 2000, 14).

The slurry pipeline, transports its coal-water mix off the reservation, where it is converted into electricity for the use of consumers in California as well as parts of Utah and central Arizona. In the meantime, many of the Dineh (traditional Navajo) who live at Black Mesa and in most other areas of the reservation have no electricity.

When Peabody was confronted during the late 1990s with complaints that it wasn't paying enough for its coal, the Hopi Nation government renegotiated to increase the company's payments by 10 percent, or about $1 million more per year. The deal was sweetened by a $1 million bonus paid at the signing of the agreement. By 1998, coal revenues accounted for nearly 40 percent of the Navajo Nation's governmental budget and 80 percent of the Hopi Nation's budget ("Figure This" 1998). One critic asked, "Will this increase help the 50 percent unemployed Navajo out on the Rez, and will it raise the yearly average income of $750? Since the wells under the Rez are almost dry, from whence will come the water needed to move this coal to Albuquerque and Nevada?" ("Figure This" 1998).

Many Navajo and Hopi have moved from the path of the growing Peabody Mine, accepting the company's offer of new homes with plumbing and solar power. Others, however, have decided to remain in their traditional homes. Maxine Kescoli is one of a number of elders who have refused to move to make way for expansion of the coal mine. Peabody has mining rights to the land under her traditional Navajo hogan, which is one mile from the Kayenta Mine. "My umbilical cord is buried here," Kescoli said ("Figure This" 1998).

To illustrate why she prefers to remain in her traditional home, Kescoli called on *hozho*, the Navajo foundation of belief, "to walk in beauty." This belief runs counter to exploitation of coal and other resources. This clash is evident not only at Black Mesa but also in many areas of the Navajo Nation where coal is mined. Emma Yazzie, another elder, herded sheep and goats in the shadow of the Four

Corners strip mine draglines until she died, even as pollution fouled her home. She walked to the bottom of strip mines near her home and impeded the draglines—a one-woman, anti-mining protest movement (Johansen and Maestas 1979, 143–46).

The clash of cultures is very sharp:

> During the past 30 years, 10,000 Navajo sheepherders have been removed from land near Black Mesa, some at gunpoint. Four thousand graves and sacred places have been desecrated. Charges are that the Peabody Coal Company, U.S. agents, and Hopi police have impounded livestock illegally. . . . All of the wells may be dry within four years. Peabody Coal plants trees as "restoration." All of the trees are dead. This is not "walking in beauty." ("Figure This" 1998)

Before the mid-1970s, Black Mesa was home to several thousand Navajo and Hopi sheepherders, weavers, silversmiths, and farmers whose families had lived there for several hundred years. Over time, many of these people were forced off Black Mesa by the encroaching strip mine. During the late 1980s, a UN report described the case of the forced relocation as "one of the most flagrant violations of indigenous peoples' human rights in this hemisphere" ("A Brief History" n.d.).

The scope of Peabody's operation at Black Mesa can be sketched using statistics. For example, in 1998, Peabody Western Coal Company (a subsidiary of the Peabody Group) removed just under 11.8 million tons of coal from its Arizona mines. Since mining began on Black Mesa three decades ago, close to 40 billion gallons of groundwater have been pumped to feed Peabody's coal-slurry pipeline. With the pipeline conveying as much as 43,000 tons of slurried coal per day, the company pumps as much as 120,000 gallons of water per hour ("Clean Coal's" n.d.). The coal slurry competes with local residents for the only source of water for the Hopi and Western Navajo. By the late 1990s, Peabody's strip mine at Black Mesa had expanded to 100 square miles, "the largest privately-owned coal mine in the world" ("A Brief History" n.d.). Into this pit have fallen burial and sacred sites, religious structures, and Anasazi ruins.

The U.S. federal government has relocated 9,000 Navajo to "New Lands" at Sanders, Arizona, on land contaminated by the largest radioactive waste spill in North America, in and near the Rio Puerco. According to one source, "Some people living there have died from cancer or are dying from it now. The birth defect rate is outstanding. . . . The suicide rate is outstanding as well" ("A Brief History" n.d.). A radiation meter outside a school on the relocation lands registers 700 rads. The relocation agents have resisted complaints that the "new lands" are contaminated by saying that people will be unaffected if they do not drink the water. There is, however, no other water source in the area. People live in trailers, their traditional economy and way of life destroyed.

Other families of Navajo refugees from Black Mesa have drifted from place to place for many years. Some live in shacks, others live in vehicles, while the lucky ones squeeze in with other family members ("A Brief History" n.d.):

> Others [have] found themselves having to pay for water, heat, food, electricity, [and] taxes, things they never had to deal with before. Many of the elders speak little or no English—people who had no experience with a cash economy have been moved to border towns. These Navajos were warehoused in substandard housing. . . . While the relocation law required the federal government to provide community facilities and services and to minimize the adverse social, cultural, and economic effects of relocation, that promise remains unfulfilled almost two decades later. Many find it impossible to get jobs, and they are forced into homelessness. The genocide is complete. ("A Brief History" n.d.)

Iyawata Britten Schneider (n.d.) wrote in *Country Road News* that "in Arizona, the Department of Interior and the Bureau of Indian Affairs created a land dispute between the Hopi and Dineh people by acquiring land and paying the Dineh large sums and the Hopi next to nothing. This land dispute was then used to support the passage of the Relocation Act of 1974 which forced the move of 10,000 Dineh to 'new lands.' This relocation brought with it increased health risks, high suicide rates, and severe depression among the Dineh. The movement of the Dineh allowed for the leasing of more land for the coal mines."

Coal mining is antithetical to traditional herding and agriculture in the region. According to Schneider (n.d.),

> Because the southwest is arid and has high winds, coal dust can be seen on the sheep herds, on the land, in the water supply and the air. Residents of the coal mining areas suffer from chronic lung disease and high cancer rates. Livestock drinking water from nearby ponds die within a few hours and traditional crops fail. In its reparation of land, Peabody does not separate the topsoil from the bottom, leaving the soil with high saline content. Vegetation has not survived in this soil.

The Black Mesa coal-slurry pipeline also is prone to occasional spills, such as one that occurred over two and a half years between 1999 and 2001. Black Mesa Pipeline Inc. agreed to pay penalties of $128,000 for discharging almost 485,000 gallons of coal slurry into northern Arizona. The Arizona Department of Environmental Quality discovered the violations during a series of inspections of Black Mesa's facilities.

"Had the pipeline been properly maintained, these spills would not have occurred," said Alexis Strauss, water division director for the U.EPA's Pacific Southwest Office. "Desert ecosystems are quite fragile and filling arroyos with crushed coal is unnecessary and unacceptable" ("Black Mesa Spill" 2001). Corrosion of the pipeline sometimes results in ruptures; the slurry can harm local wildlife

Increasing numbers of Hopis and Navajos have been protesting the fact that Peabody's use of local water is destroying their way of life piecemeal. Former Hopi Chairman Ferrell Secakuku joined Navajos from Big Mountain during

2001 to protest the use of aquifer water for the Black Mesa coal slurry and urge the creation of new, sustainable forms of energy. "We found out the water table is being depleted," Secakuku said, denying reports by the federal government and the Peabody Western Coal Company that have claimed otherwise. Secakuku spoke to a crowd of American Indians and other environmentalists that included longtime opponents of Navajo relocation at Big Mountain. Secakuku and other protesters entered the offices of Black Mesa Pipeline Inc. to urge officials to halt the slurry pipeline that annually depletes 1.3 billion gallons from the Navajo Aquifer. He said using water to transport coal threatens to leave the Hopi village of Moenkopi without water by 2011. "Every time you breathe, Peabody is pumping 50 gallons," Secakuku told the crowd. During the protest, Roberta Blackgoat held a sign proclaiming, "The Creator Is the Only One Who Is Going to Relocate Me" (Norrell 2001).

Secakuku added that there has been no response from the Department of the Interior to a proposal that would provide Lake Powell water to Hopi for drinking purposes. He said that if no alternative water source is provided by 2004, he and Bucky Preston would march to the Hopi Tribal Council and demand termination of contracts with Peabody, ending coal mining on Black Mesa. "By the year 2004, we will stop the mining," he said. "Soon the aquifer will crumble and no longer take the recharge" (Norrell 2001).

In defense of the mine, Peabody Coal's John Wasik, executive for Southwestern Operations, said, "Continued operation of the Black Mesa Mine is in the public interest and will provide long-term economic benefits to the Hopi Tribe and the Navajo Nation. It also provides an essential and secure energy supply for more than 1 million Southwest families who rely on [it for] electricity." Peabody further maintains that the mine "will inject about $1.5 billion in direct economic benefits into reservation economies in royalties, taxes, wages, and vendor contracts over the proposed extension [of mine operations]" for 15 years, from 2005 to 2020. The mine employed about 250 people in 2002; 96 percent of the mine's workforce and 82 percent of its supervisory staff were Native American, according to Peabody (Wanamaker 2002).

In recent years, concerns about the Black Mesa strip mine's effects on local land and people have not ebbed. Peabody has sought to expand its operations on roughly 100 square miles of Hopi and Navajo lands, using aquifers from sacred springs that provide drinking water for local communities. "The mining-related impacts on the aquifer are more significant than have been recognized or acknowledged," said Dr. Daniel Higgins, who performed the analysis as part of Arid Lands Resource Studies, graduate interdisciplinary programs, University of Arizona, Tucson. "Despite what these models predicted years ago, I think any reasonable person who looks at the data would conclude that the rates of water level decline at Kayenta and spring discharge decline at Moenkopi are directly related to Peabody's groundwater withdrawals," Higgins was quoted as saying in the organization's press release (Berry 2011b).

Several Hopi and Navajo organizations, including, Dine' CARE (Citizens Against Ruining Our Environment), Black Mesa Water Coalition, and To'Nizhoni Ani

("Beautiful Water Speaks"), agreed with these findings, as did the Sierra Club and the Center for Biological Diversity. The report was released at a "critical time while OSM (Office of Surface Mining Reclamation and Enforcement) is preparing an environmental assessment to analyze the impacts of the Kayenta Mine," Nicole Horseherder of To' Nizhoni Ani said. "OSM officials now need to address and respond to this report before they let Peabody off the hook for damage to the Navajo aquifer (N-Aquifer)" (Berry 2011b).

## *MERCURY AND COAL POWER ON THE NAVAJO NATION*

Some changes have come at Navajo. The EPA has limited emissions of nitrous oxides, including the ones from the Four Corners Power Plant that were poisoning local residents during the 1970s. The EPA in 2011 also issued tough limits on mercury, arsenic, acid gas, and other toxic substances. Even with recent restrictions on its pollution, coal-generated electric power "represent America's single biggest source of air pollution, affecting our waterways, destroying ecosystems, and polluting the air we breathe," said Environment America. "Pollution from coal-fired power plants in particular contributes to four of the five leading causes of mortality in the United States." The Navajo Nation has a lease on the Four Corners Power Plant near Farmington, New Mexico, until 2041; it is said to provide $7 million a year in economic benefits, including jobs (Berry 2011a).

Evidence of mercury toxicity has become widespread around the three coal-fired power plants on Navajo lands. "Teachers are telling us there's a lot of developmental problems for the kids—that's mercury," said Lori Goodman, co-ordinator for Dine' CARE, describing problems at the school nearest the Four Corners Power Plant. "And the autism—that's mercury-related" (Berry 2011a). Exposure to mercury, even at low levels, affects children permanently, leading to lower IQs, attention-deficit problems, lower verbal skills, and motor-control problems. Exposure in utero to high levels is associated with blindness, mental retardation, and seizures.

In the midst of the controversy over mercury and other emissions from Navajo coal plants, a parody Internet site for Peabody Energy appeared on the Internet titled "Coal: It's the Safest Energy There Is":

> The satirical site (www.CoalCares.com), launched in early May by the organization Coal Kills Kids, appears to offer free "Puff-Puff" inhalers to families who live within 200 miles of a coal plant, reported the *Los Angeles Times*, along with coupons toward the purchase of asthma medication. There's even a F.A.Q. and Media page, plus fun games in the "Kidz Koal Korner." (Cevallos 2011)

In the face of activist satire, however, the Navajo and Hopi governments cautioned that closure of coal-fired power plants could cost $40 million in revenue and as many as 1,000 jobs ("Cleaning Up King Coal?" 2011).

## FURTHER READING

Abel, Heather. 1997. "The Rise and Fall of a Gold Mining Company." *High Country News* 29, no. 24 (December 22). http://www.hcn.org/servlets/hcn.Article?article_id=3860

"Anglo American Withdraws from Pebble Mine." 2013. Environment News Service, September 20. http://ens-newswire.com/2013/09/20/anglo-american-withdraws-from-alaskas-pebble-mine

Ball, David P. 2012. "Tsilhqot'in Vow to 'Stand Ground' against New Prosperity Mine at Taseko AGM." Indian Country Today Media Network, June 7. http://indiancountrytodaymedianetwork.com/article/tsilhqotin-vow-to-stand-ground-against-new-prosperity-mine-at-taseko-agm-116686

Berry, Carol. 2011a. "Coal Power Plants and Mercury on the Navajo Nation." Indian Country Today Media Network, August 8. http://indiancountrytodaymedianetwork.com/article/coal-power-plants-and-mercury-on-the-navajo-nation-46166

Berry, Carol. 2011b. "Navajo Aquifer Concerns May Prove True." Indian Country Today Media Network, August 5. http://indiancountrytodaymedianetwork.com/article/navajo-aquifer-concerns-may-prove-true-46043

"Black Mesa Spill Nets $128,000 Fine." 2001. Arizona Water Resource: News Briefs, June. http://ag.arizona.edu/AZWATER/awr/mayjune01/news.html

"A Brief History of Relocation on Black Mesa." N.d. http://www.blackmesais.org/cultural_sen_history.html

"Bristol Bay Tribes' Fight to Fend Off Pebble Mine Highlighted in *National Geographic*." 2012. Indian Country Today Media Network, November 19. http://indiancountrytodaymedianetwork.com/bristol-bay-tribes-fight-to-fend-off-pebble-mine-highlighted-in-national-geographic

Cevallos, Marissa. 2011. "Coal Power? It's Fine—and Have an Inhaler, Says Fake Website." *Los Angeles Times*, May 11. http://articles.latimes.com/2011/may/11/news/la-heb-coal-parody-20110511

"Cheyenne River Sioux Tribe Wins Suit against Homestake." 1999. *Drillbits and Tailings* 4, no. 11 (July 23). http://www.moles.org/ProjectUnderground/drillbits/4_11/2.html

Chythlook-Sifsof, Callan J. 2013. "Native Alaska, under Threat." *New York Times*, June 28. http://www.nytimes.com/2013/06/28/opinion/native-culture-under-threat.html

"Clean Coal's Dirty Facts." N.d. http://www.blackmesais.org/clean_coals_dirty_facts.html

"Cleaning Up King Coal? Tribes Appeal Tighter Restrictions on Power Plant Emissions." 2011. Indian Country Today Media Network, May 26. http://indiancountrytodaymedianetwork.com/article/cleaning-up-king-coal%3F-tribes-appeal-tighter-restrictions-on-power-plant-emissions-35965

Cohen, Faye G. 1986. *Treaties on Trial: The Continuing Controversy over Northwest Fishing Rights*. Seattle: University of Washington Press.

"Company That Designed Burst B.C. Tailings Pond Was Hired by Pebble Mine in Bristol Bay." 2014. Indian Country Today Media Network, August 8. http://indiancountrytodaymedianetwork.com/2014/08/08/company-designed-burst-bc-tailings-pond-was-hired-pebble-mine-bristol-bay-156315

Cortes, Amber. N.d. "Treaty Rights and Totem Poles: How One Tribe Is Carving Out a Resistance to Coal." *Grist: A Beacon in the Smog*. http://exp.grist.org/lummi

Dakotah. 2013. "Taȟċa Wakutėpi: Where They Killed Deer." Indian Country Today Media Network, January 19. http://indiancountrytodaymedianetwork.com/2013/01/19/ta%C4%A5%C4%8B-wakut%C4%97pi-where-they-killed-deer-147081

"Dam Breach at Mount Polley Mine in British Columbia." 2014. NASA Earth Observatory, August 17. http://earthobservatory.nasa.gov/IOTD/view.php?id=84202&src=eoa-iotd

Emmons, Della Gould. 1965. *Leschi of the Nisquallies*. Minneapolis: T. S. Denison.

Fernandez, Lydia. 1998. "Gold Mine Threatens Quechan Sites." *Native Americas* 15, no. 2 (Summer): 6.

"Figure This One!" 1998. *Four Corners Clamor* 1, no. 6 (Summer/Fall). http://www.ausbcomp.com/redman/clamor6.htm#newdeal

"$4.3 Million Settlement over Reid-Gardner Coal Plant for Moapa Paiute." 2015. Indian Country Today Media Network, August 6. http://indiancountrytodaymedianetwork.com/2015/08/06/43-million-settlement-over-reid-gardner-coal-plant-moapa-paiute-161299

Gedicks, Al. 1993. *The New Resource Wars: Native and Environmental Struggles against Multinational Corporations*. Boston: South End Press.

Giuliano, Jackie Alan. 2001. "Killing Tomorrow for a Few Megawatts Today." Environment News Service, March. http://ens.lycos.com/ens/mar2001/2001L-03-30g.html

Hansen, Terri. 2012. "Northwest Tribes Fight for Treaty Rights in Face of Coal-Transport Plan." Indian Country Today Media Network, August 15. http://indiancountrytodaymedianetwork.com/article/northwest-tribes-fight-for-treaty-rights-in-face-of-coal-transport-plan-129453

Hernandez, Ruben. 2012. "An Energy Revolution on the Navajo Reservation." New America Media, December 10. http://newamericamedia.org/2012/12/an-energy-revolution-on-the-navajo-reservation.php

"Horrific Toxic Spill in B.C. Called Another Exxon Valdez." 2014. Indian Country Today Media Network, August 7. http://www.nativenewstoday.com/2014/08/07/horrific-toxic-spill-in-b-c-called-another-exxon-valdez

"Hotspots: California." 2001. *Drillbits and Tailings* 6, no. 9 (November 30). http://www.moles.org

Huff, Andrew. 2000. "Gold Mining Threatens Communities." The Progressive Media Project, July 11. http://www.progressive.org/mpdvah00.htm

Hughes, Pam. 2013a. "Bad River Band Establishes Legal Defense Fund to Stop Proposed Iron-Ore Mine." Indian Country Today Media Network, March 19. http://indiancountrytodaymedianetwork.com/2013/03/19/bad-river-band-establishes-legal-defense-fund-stop-proposed-iron-ore-mine-148251

Hughes, Pam. 2013b. "Wisconsin Tribes: What Part of 'No Mining!' Don't You Understand?" Indian Country Today Media Network, July 7. http://indiancountrytodaymedianetwork.com/2013/07/07/wisconsin-tribes-what-part-no-mining-dont-you-understand-150309

James, Jewell. 2014. "2014 Totem Pole Journey Honors Tribes' Stewardship of Land, Water." *Bellingham Herald*, August 11. http://www.bellinghamherald.com/2014/08/11/3792147_2014-totem-pole-journey-honors.html

Johansen, Bruce E. 1998. "Exxon vs. Chippewas in Wisconsin." *Native Americas* 15, no. 1 (Spring): 6.

Johansen, Bruce, and Roberto Maestas. 1979. *Wasi'chu: The Continuing Indian Wars*. New York: Monthly Review Press.

Kaufman, Dan. 2014. "The Fight for Wisconsin's Soul." *New York Times*, March 29. http://www.nytimes.com/2014/03/30/opinion/sunday/the-fight-for-wisconsins-soul.html

Landry, Alysa. 2014. "Is the Navajo Mine a Viable Option? Growing Opposition Doesn't Think So." Indian Country Today Media Network, February 6. http://indiancountrytodaymedianetwork.com/2014/02/06/navajo-mine-viable-option-growing-opposition-doesnt-think-so-153407

Lazenby, Henry. 2014. "Taseko Stocks Tumble as Fed Rejects New Prosperity, BC—Again." *Mining Weekly*, February 27. http://www.miningweekly.com/article/taseko-stocks-down-as-fed-rejects-new-prosperity-bc-again-2014-02-27

Lee, Tanya. 2013. "Important Cultural, Religious and Historical Resources Threatened by Drilling." Indian Country Today Media Network, April 8. http://indiancountrytodaymedianetwork.com/2013/04/08/important-cultural-religious-and-historical-resources-threatened-drilling-148622

"Lummi Nation Officially Opposes Coal Export Terminal in Letter to Army Corps of Engineers." 2012. Indian Country Today Media Network, August 2. http://indiancountrytodaymedianetwork.com/2013/08/02/lummi-nation-officially-opposes-coal-export-terminal-letter-army-corps-engineers-150718

"Major Victory: Canadian Supreme Court Hands Tsilhqot'in Aboriginal Title." 2014. Indian Country Today Media Network, June 26. https://indiancountrytodaymedianetwork.com/2014/06/26/major-victory-canadian-supreme-court-hands-tsilhqotin-aboriginal-title-155493

McNeel, Jack. 2012. "Railroad to Disaster? Inland Tribes Fight to Avert Coal-Train Destruction." Indian Country Today Media Network, December 17. http://indiancountrytodaymedianetwork.com/article/railroad-disaster-inland-tribes-fight-avert-coal-train-destruction-146338

Millet, Lydia. 2015. "Selling Off the Holy Land." *New York Times*, May 29, A-23, A-27.

Minard, Anne. 2013. "Arizona Tribes and Environmental Groups Oppose HR 287 for Mining in the Tonto National Forest." Indian Country Today Media Network, February 26. http://indiancountrytodaymedianetwork.com/2013/02/26/arizona-tribes-and-environmental-groups-oppose-hr-287-mining-tonto-national-forest-147894

"Moapa Paiute to Host Second Solar Project on Its Lands." 2014. Indian Country Today Media Network, May 29. http://indiancountrytodaymedianetwork.com/2014/05/29/moapa-paiute-host-2nd-solar-project-its-lands-155071

"Moapa Paiute March 50 Miles in Anti-Coal Protest." 2012. Indian Country Today Media Network, April 27. http://indiancountrytodaymedianetwork.com/article/moapa-paiute-march-50-miles-in-anti-coal-protest-110450

"Moapa Paiute Sue over Coal Plant Contaminants." 2013. Indian Country Today Media Network, August 9. http://indiancountrytodaymedianetwork.com/2013/08/09/moapa-paiute-sue-over-coal-plant-contaminants-150806

"National Parks Conservation Group Objects to Navajo Generating Station Proposal." 2013. Indian Country Today Media Network, November 13. http://indiancountrytodaymedianetwork.com/2013/11/13/national-parks-conservation-group-objects-navajo-generating-station-proposal-152238

"Native Alaskans Laud Environmental Protection Agency's Nixing of Pebble Mine in Bristol Bay." 2014. Indian Country Today Media Network, March 3. http://indiancountrytodaymedianetwork.com/2014/03/03/native-alaskans-laud-environmental-protection-agencys-nixing-pebble-mine-bristol-bay

"Navajo Activists Spearhead Energy Transformation in the Four Corners Area." 2012. Indian Country Today Media Network, December 11. http://indiancountrytoday

medianetwork.com/article/navajo-activists-spearhead-energy-transformation-four-corners-area-146246

Norrell, Brenda. 2001. "Hopi and Navajo Protest Pipeline." *Indian Country Today Global Report* 129 (July 5–11).http://www.agrnews.org/issues/129/nationalnews.html

"Northwest Tribes Step Up Opposition to Proposed Coal Terminals." 2012. Indian Country Today Media Network, October 19. http://indiancountrytodaymedianetwork.com/article/northwest-tribes-step-up-opposition-to-proposed-coal-terminals-140853

"Obama Declares Bristol Bay Off Limits to New Oil and Gas Drilling Leases." 2014. Indian Country Today Media Network, December 17. http://indiancountrytodaymedianetwork.com/2014/12/17/obama-declares-bristol-bay-limits-new-oil-and-gas-drilling-leases-158329

"PBS News Hour: Lummi Fighting Cherry Point Coal Export Terminal." 2012. Indian Country Today Media Network, August 4. http://indiancountrytodaymedianetwork.com/2013/08/04/pbs-newshour-highlights-lummi-battle-over-cherry-point-coal-export-terminal-150726

Pember, Mary Annette. 2013a. "Automatic Weapons & Guards in Camo: Welcome to Mining Country, Wis." Indian Country Today Media Network, July 9. http://indiancountrytodaymedianetwork.com/2013/07/09/automatic-weapons-guards-camo-welcome-mining-country-wis-150342

Pember, Mary Annette. 2013b. "County Declares Harvest Camp Illegal, Moves to Shut It Down." Indian Country Today Media Network, July 23. http://indiancountrytodaymedianetwork.com/story/wis-mining-war

Pember, Mary Annette. 2013c. "Racism & Violent Threats: Wis. Mining War Gets Uglier and Scarier." Indian Country Today Media Network, July 16. http://indiancountrytodaymedianetwork.com/2013/07/16/wis-mining-racism-and- threats-reach-facebook-150449

Pember, Mary Annette. 2013d. "Republican Favoritism of GTAC Gets Public Comment Session." Indian Country Today Media Network, September 3. http://indiancountrytodaymedianetwork.com/2013/09/03/republican-favoritism-gtac-gets-public-comment-session-151137

Pember, Mary Annette. 2013e. "Who Is Illegal in Wis. Mining Country? Harvest Camp or GTAC?" Indian Country Today Media Network, July 20. http://indiancountrytodaymedianetwork.com/2013/07/20/who-illegal-wis-mining-country-harvest-camp-or-gtac-150512

Pember, Mary Annette. 2005. "A Bad River Win; Gogebic Taconite Putting Wisc. Mine on Hold." Indian Country Today Media Network, March 3. http://indiancountrytodaymedianetwork.com/2015/03/03/bad-river-win-gogebic-taconite-putting-wisc-mine-hold-159434

Pierpoint, Mary. 2001. "Reno-Sparks Fights Kitty Litter Mine; High Traffic, Dust, Potentially Toxic Minerals All Concerns." *Indian Country Today*, December 12, A-1.

Pogash, Carol. 2013. "Governor, Chippewas Battle over Mine." *USA Today*, September 9, 4-A.

"Pour Him a Tall One! Mining Exec Insists He'd Drink Water from Tailings Pond." 2014. Indian Country Today Media Network, August 7. http://indiancountrytodaymedianetwork.com/2014/08/07/pour-him-tall-one-mining-exec-insists-hed-drink-water-tailings-pond-156294

"Quapaw Take Lead in Tar Creek Superfund Site Cleanup in Oklahoma, Kansas." 2014. Indian Country Today Media Network, June 4. http://indiancountrytoday

medianetwork.com/2014/06/04/quapaw-take-lead-tar-creek-superfund-site-cleanup-oklahoma-kansas-155155

Riordan, Michael. 2013. "Don't Sell Cheap U.S. Coal to Asia." *New York Times*, February 13, A-23. http://www.nytimes.com/2014/02/13/opinion/dont-sell-cheap-us-coal-to-asia.html

Schneider, Andrew. 2001. " 'A Wounded Mountain Spewing Poison': The Mining of the West: Profit and Pollution on Public Lands: A *Post-Intelligencer* Special Report." *Seattle Post-Intelligencer*, June 12. http://seattlepi.nwsource.com/specials/mining/27076_lodgepole12.shtml

Schneider, Iyawata Britten. N.d. "Environmental and Human Rights Devastation in the Southwest." *Country Road News*. http://www.countryroadchronicles.com/Articles/CountryRoadNews/devastation.html

Schneider, Keith. 1990. "Washington Nuclear Plant Poses Risk for Indians." *New York Times*, September 3, A-9.

Smitten/Vimeo, Susan. 2014. "First Nations Exult as Canadian Government Rejects Taseko's New Prosperity Mine." Indian Country Today Media Network, February 27. http://indiancountrytodaymedianetwork.com/2014/02/27/first-nations-exult-canadian-government-rejects-tasekos-new-prosperity-mine-153777

"360 Scientists Urge Environmental Protection Agency to Quash Bristol Bay Pebble Mine." 2014. Indian Country Today Media Network, February14. http://indiancountrytodaymedianetwork.com/2014/02/14/360-scientists-urge-environmental-protection-agency-quash-bristol-bay-pebble-mine-153559

Trahant, Mark. 2012. "Elections 2012: Vote for Coal? Northwest Tribes Raise New Concerns." Indian Country Today Media Network, October 1. http://indiancountrytodaymedianetwork.com/article/elections-2012%3A-vote-for-coal%3F-northwest-tribes-raise-new-concerns-137150

"Treaty Victory as Northwest Tribes Celebrate Oregon Coal Train Rejection." 2014. Indian Country Today Media Network, August 21. http://indiancountrytodaymedianetwork.com/2014/08/21/treaty-victory-northwest-tribes-celebrate-oregon-coal-train-rejection-156520

"Tsilhqot'in Acquiesce to Test Drilling, Still Determined to Stop New Prosperity Mine." 2012. Indian Country Today Media Network, March 2. http://indiancountrytodaymedianetwork.com/article/tsilhqotin-acquiesce-to-test-drilling-still-determined-to-stop-new-prosperity-mine-101023

"Uranium/Nuclear Issues and Native Communities." 2001. Indigenous Environmental Network. http://www.ienearth.org/nuciss.html

Vandevelder, Paul. 2000. "Between a Rock and a Dry Place." *Native Americas* 17, no. 2 (Summer): 10–15.

Vandevelder, Paul. 2001. "A Native Sense of Earth." *Native Americas* 18, no. 1 (Spring): 42–49.

Walker, Richard. 2014. "Lummi Totem Pole Journey Rallies Voices against Environmental Destruction." Indian Country Today Media Network, September 2. http://indiancountrytodaymedianetwork.com/2014/09/02/lummi-totem-pole-journey-rallies-voices-against-environmental-destruction-156696

Walker, Richard. 2015. "Lummi Call Coal Terminals an Absolute No-Go, Invoking Treaty Rights." Indian Country Today Media Network, February 26. http://indiancountrytodaymedianetwork.com/2015/02/26/lummi-call-coal-terminals-absolute-no-go-invoking-treaty-rights-159381

Wanamaker, Tom. 2002. "Hopi Tribe Opposes Lease Extension on Water-Thirsty Black Mesa Mine." *Indian Country Today*, February 13, C-1.

Warrick, Joby. 2015. "Gold vs. Salmon: EPA Becomes Target by Employing Rare Preemptive 'Veto.' " *Washington Post*, February 15. http://www.washingtonpost.com/national/health-science/internal-memos-spur-accusations-of-bias-as-epa-moves-to-block-gold-mine/2015/02/15/3ff101c0-b2ba-11e4-854b-a38d13486ba1_story.html

Weaver, Jace, ed. 1996. *Defending Mother Earth: Native American Perspectives on Environmental Justice.* New York: Maryknoll.

Wilson, Reid. 2014. "Jewell 'Profoundly' Disappointed by Land Exchange at Sacred Native American Sites." *Washington Post*, December 8. http://www.washingtonpost.com/blogs/govbeat/wp/2014/12/06/jewell-profoundly-disappointed-by-land-exchange-at-sacred-native-american-sites

Woodard, Stephanie. 2012. "Moapa Paiutes Find Solar Solution Amid Coal Ash Plague." Indian Country Today Media Network, August 17. http://indiancountrytodaymedianetwork.com/article/moapa-paiutes-find-solar-solution-amid-coal-ash-plague-129554

Zhorov, Irina. 2012. "Reopening Silver Mine Could Save Town, but Destroy the Sacred Site of the Huicholes." Indian Country Today Media Network, August 7. http://indiancountrytodaymedianetwork.com/article/reopening-silver-mine-could-save-town,-but-destroy-the-sacred-site-of-the-huicholes-127755

# Selected Bibliography

Abel, Heather. 1997. "The Rise and Fall of a Gold Mining Company." *High Country News* 29, no. 24 (December 22). http://www.hcn.org/servlets/hcn.Article?article_id=3860

Allen, Lee. 2012. "Southwest Tribes Struggle with Climate Change Fallout." Indian Country Today Media Network, June 14. http://indiancountrytodaymedianetwork.com/article/southwest-tribes-struggle-with-climate-change-fallout-118386more-beavers-defecating-and-disease

Althaus, Dudley. 2001. "The Fated Forest: Nature's Way; Upsetting the Balance; Deforestation Threatens Thousands of Plant and Animal Species as Well as the Earth's Atmosphere. Studies Show Rain Forests Are Declining at Alarming Rates." *Houston Chronicle*, September 30, 6.

"American Indians Feel the Effects of Climate Change at Higher Rate Than Other Groups." 2011. Indian Country Today Media Network, August 9. http://indiancountrytodaymedianetwork.com/article/american-indians-feel-the-effects-of-climate-change-at-higher-rate-than-other-groups-46365

Amundsen, Roald. 1908. *The North West Passage*. London: E. P. Dutton & Co.

Anderson, Rebecca K., Gifford H. Miller, Jason P. Briner, Nathaniel A. Lifton, and Stephen B. DeVogel. 2008. "A Millennial Perspective on Arctic Warming from 14-C in Quartz and Plants Emerging from Beneath Ice Caps." *Geophysical Research Letters* 35: L01502. doi:10.1029/2007/GL032057

"Anglo American Withdraws from Pebble Mine." 2013. Environment News Service, September 20. http://ens-newswire.com/2013/09/20/anglo-american-withdraws-from-alaskas-pebble-mine

"Anishinaabe and Lakota Riders Protest Pipelines, on Horseback." 2013. Indian Country Today Media Network, October 18. http://indiancountrytodaymedianetwork.com/gallery/photo/anishinaabe-and-lakota-riders-protest-pipelines-horseback-151792

Anquoe, Bunty. 1993. "House Begins Investigating Possible Radiation Exposure." *Indian Country Today*, June 9, A-8.

Antonowicz, Anton. 2000. "Baking Alaska: As World Leaders Bicker, Global Warming Is Killing a Way of Life." *Daily Mirror*, November 28, 8–9.

"Arctic Ice Melting Rapidly, Study Says." 2006. *New York Times*, September 14. http://www.nytimes.com/aponline/us/AP-Warming-Sea-Ice.html

"Arctic Ice Retreating 30 Years Ahead of Projections." 2007. Environment News Service, April 30. http://www.ens-newswire.com/ens/apr2007/2007-04-30-04.asp

Arnold, Carrie. 2014. "Once upon a Mine: The Legacy of Uranium on the Navajo Nation." *Environmental Health Perspectives*. http://ehp.niehs.nih.gov/122-A44

"As Arctic Sea Ice Melts, Experts Expect New Low." 2008. *New York Times*, August 28. http://www.nytimes.com/2008/08/28/science/earth/28seaice.html

"Athabasca Oil Sands." 2011. NASA Earth Observatory, November 30. http://earthobservatory.nasa.gov/IOTD/view.php?id=76559&src=eoa-iotd

"Athabasca Oil Sands in All Their Terrible Glory." 2011. Indian Country Today Media Network, September 23. http://indiancountrytodaymedianetwork.com/department/environment?page=13

Austen, Ian. 2013. "Canadian Review Panel Approves Plans for an Oil Pipeline." *New York Times*, December 19. http://www.nytimes.com/2013/12/20/business/international/canadian-review-panel-approves-plans-for-an-oil-pipeline.html

Authier, Philip. "Cree Pact Clears Final Hurdle." *Montreal Gazette*, February 5, 2002. http://www.canada.com/montreal/montrealgazette/story.asp?id={43F9B38E-7577-4579-AEED-90A8F9638936}

Baker, James N. 1990. "Keeping a Deadly Secret: The Feds Knew the Mines Were Radioactive." *Newsweek*, June 18, n.p.

Ball, David P. 2012a. "Athabasca Chipewyan Launch Treaty 8 Challenge to Shell Canada over Oil Sands." Indian Country Today Media Network, October 5. http://indiancountrytodaymedianetwork.com/article/athabasca-chipewyan-launch-treaty-8-challenge-to-shell-canada-over-oil-sands-137632

Ball, David P. 2012b. "Spiritual Leaders Vow to Defend Mother Earth from Oil Sands and Pipelines with Spiritual Declaration." Indian Country Today Media Network, November 23. http://indiancountrytodaymedianetwork.com/spiritual-leaders-vow-to-defend-mother-earth-from-oil-sands-and-pipelines-with-spiritual-declaration-147332

Ball, David P. 2012c. "U.S. and Canada-Wide Protests Target Pacific Trails' Proposed Fracking Pipeline." Indian Country Today Media Network, November 27. http://indiancountrytodaymedianetwork.com/article/us-and-canada-wide-protests-target-pacific-trails-proposed-fracking-pipeline-145895

Ball, David P. 2012d. "Tsilhqot'in Vow to 'Stand Ground' against New Prosperity Mine at Taseko AGM." Indian Country Today Media Network, June 7. http://indiancountrytodaymedianetwork.com/article/tsilhqotin-vow-to-stand-ground-against-new-prosperity-mine-at-taseko-agm-116686

Ball, David P. 2013a. "Enbridge Faces Rising Opposition to Northern Gateway Pipeline with Protests, Arrests." Indian Country Today Media Network, January 25. http://indiancountrytodaymedianetwork.com/2013/04/04/red-lake-band-chippewa-determined-blockade-atop-four-enbridge-pipelines-minnesota-148600

Ball, David P. 2013b. Fracking Troubles Atlantic First Nations after Two Dozen Protesters Arrested. Indian Country Today Media Network, June 28. http://indiancountrytodaymedianetwork.com/2013/06/28/fracking-troubles-atlantic-first-nations-after-two-dozen-protesters-arrested-150192

Ball, David. 2014a. "Neil Young: Blood of First Nations People Is on Canada's Hands." Indian Country Today Media Network, January 14. http://indiancountrytoday

medianetwork.com/2014/01/14/neil-young-blood-first-nations-people-canadas-hands-153104

Ball, David P. 2014b. "Tensions Simmer after Kinder Morgan Confrontation over Pipeline Extension in Burnaby, B.C." Indian Country Today Media Network, December 12. http://indiancountrytodaymedianetwork.com/2014/12/12/tensions-simmer-after-kinder-morgan-confrontation-over-pipeline-extension-burnaby-bc

Barreda, Andres. "Militarization and Oil in Chiapas." *La Jornada*, August 17, n.p.

Barreiro, Jose, ed. 1991. "Readings on James Bay." *Akwe:kon Journal* 8, no. 4 (Winter): n.p.

Barry, Tom, and Beth Wood 1978. "Uranium on the Checkerboard: Crisis at Checkpoint." *American Indian Journal*, June, n.p.

Berry, Carol. 2011a. "Coal Power Plants and Mercury on the Navajo Nation." Indian Country Today Media Network, August 8. http://indiancountrytodaymedianetwork.com/article/coal-power-plants-and-mercury-on-the-navajo-nation-46166

Berry, Carol. 2011b. "Navajo Aquifer Concerns May Prove True." Indian Country Today Media Network, August 5. http://indiancountrytodaymedianetwork.com/article/navajo-aquifer-concerns-may-prove-true-46043

Berry, Carol. 2011c. "Scholar Daniel Wildcat in Discussing the Environment: It's Relatives, Not 'Resources.'" Indian Country Today Media Network, October 1. http://indiancountrytodaymedianetwork.com/gallery/photo/scholar-daniel-wildcat-in-discussing-the-environment%3A-it%E2%80%99s-relatives%2C-not-%E2%80%98resources%E2%80%99-56599

Berry, Carol. 2012a. "Alberta Oil Sands Up Close: Gunshot Sounds, Dead Birds, a Moonscape." Indian Country Today Media Network, February 2. http://indiancountrytodaymedianetwork.com/article/alberta-oil-sands-up-close%3A-gunshot-sounds%2C-dead-birds%2C-a-moonscape-95444

Berry, Carol. 2012b. "Oglalas' Keystone XL Pipeline Position Issued." Indian Country Today Media Network, January 31. http://indiancountrytodaymedianetwork.com/article/oglalas%E2%80%99-keystone-xl-pipeline-position-issued-95128

Berry, Carol. 2013. "Tribal Members Sign Treaty Calling for an End to Alberta Oil Sands Development and Keystone XL." Indian Country Today Media Network, January 31. http://indiancountrytodaymedianetwork.com/2013/01/31/tribal-members-sign-treaty-calling-end-alberta-oil-sands-development-and-keystone-xl

Bienkowski, Brian. 2013. "Contaminated Tribe: Hormone-Blocking Chemicals Found in First Nation Families." Environmental Health News, November 11. http://www.environmentalhealthnews.org/ehs/news/2013/contaminated-tribe

Bigert, Claus 1995. "A People Called Empty." In *Amazon of the North: James Bay Revisited* by Rainer Wittenborn and Claus Biegert. Program for show, Santa Fe Center for Contemporary Arts, August 4–September 5, n.p.

"Black Mesa Spill Nets $128,000 Fine." 2001. Arizona Water Resource: News Briefs, June. http://ag.arizona.edu/AZWATER/awr/mayjune01/news.html

Blais, Jules M., David W. Schindler, Derek C. G. Muir, Lynda E. Kimpe, David B. Donald, and Bruno Rosenberg. 1998. "Accumulation of Persistent Organochlorine Compounds in Mountains of Western Canada." *Nature* 395 (October 8): 585.

"Blood Tribe Members Set Up Anti-Fracking Blockade." 2011. Indian Country Today Media Network, September 10. http://indiancountrytodaymedianetwork.com/article/blood-tribe-members-set-up-anti-fracking-blockade-53321

Boucher O., G. Muckle, J. L. Jacobson, R. C. Carter, M. Kaplan-Estrin, P. Ayotte, É. Dewailly, and S. W. Jacobson. 2014. "Domain-Specific Effects of Prenatal Exposure to PCBs, Mercury, and Lead on Infant Cognition: Results from the

Environmental Contaminants and Child Development Study in Nunavik." *Environmental Health Perspectives*. http://dx.doi.org/10.1289/ehp.1206323

Bowen, Jerry. 2002. "Dramatic Climate Change in Alaska." *CBS Morning News*, August 29.

Boyd, Robert S. 2002. "Earth Warming Could Open Up a Northwest Passage." *Pittsburgh Post-Gazette*, November 11, A-1.

Bradsher, Keith. 2008. "A Drought in Australia, a Global Shortage of Rice." *New York Times*, April 17. http://www.nytimes.com/2008/04/17/business/worldbusiness/17warm.html

Braem, Nicole M. 1997. "Greenpeace Activists Visit Yup'ik and Inupiat Villages in Alaska to Gather Information about Global Warming." *Nunatsiaq News*, August 8. http://www.nunanet.com/~nunat/week/70808.html

Brantley, Susan L., and Anna Meyendorff. 2013. "The Facts on Fracking." *New York Times*, March 13. http://www.nytimes.com/2013/03/14/opinion/global/the-facts-on-fracking.html

"A Brief History of Relocation on Black Mesa." N.d. http://www.blackmesais.org/cultural_sen_history.html

"Bristol Bay Tribes' Fight to Fend Off Pebble Mine Highlighted in *National Geographic*." 2012. Indian Country Today Media Network, November 19. http://indiancountrytodaymedianetwork.com/bristol-bay-tribes-fight-to-fend-off-pebble-mine-highlighted-in-national-geographic

Brooke, James. 2000. "Canada's Bleak North Is Fertile Ground for Suicide." *Canadian Aboriginal News*, December 18. http://www.canadianaboriginal.com/health/health21a.htm

Brown, DeNeen L. 2001. "Culture Corrosion in Canada's North; Forced into the Modern World, Indigenous Inuit Struggle to Cope." *Washington Post*, July 16, A-1.

Brugge, Doug, Timothy Benally, and Esther Yazzie-Lewis, eds. 2006. *The Navajo People and Uranium Mining*. Albuquerque: University of New Mexico Press.

Buxton, James. 2000. "Suspects in the Mystery of Scotland's Vanishing Salmon: Fish Farms, Seals and Global Warming Are All Blamed for What Some See as a Crisis." *Financial Times*, June 13, 11.

Cadbury, Deborah. 1997. *Altering Eden: The Feminization of Nature*. New York: St. Martin's Press.

Calamai, Peter. 2000. "Chemical Fallout Hurts Inuit Babies." *Toronto Star*, March 22. http://irptc.unep.ch/pops/newlayout/press_items.htm

"Call to Action! Rally at Cuomo's State of the State: No Fracking, Yes Renewables!" 2014. Two Row Wampum Renewal Campaign, January 14. http://honorthetworow.org/call-to-action-rally-at-cuomos-state-of-the-state-no-fracking-yes-renewables

"Canada Approves Enbridge Pipeline through B.C., First Nations Will Sue." 2014. Indian Country Today Media Network, June 19. http://indiancountrytodaymedianetwork.com/2014/06/19/canada-approves-enbridge-pipeline-through-bc-first-nations-will-sue-155377

"Canada Enbridge Approval Unites First Nations, Amnesty, Rights Groups in Opposition." 2014. Indian Country Today Media Network, June 20. http://indiancountrytodaymedianetwork.com/2014/06/20/canada-enbridge-approval-unites-first-nations-amnesty-rights-groups-opposition-155400

"Canadian Indians Paddle to New York City to Protest Quebec Power Plant." 1990. *Syracuse Post-Standard*, April 5, A-2.

"Canadian Tundra Turning Green." 2007. Environment News Service, March 3. http://www.ens-newswire.com/ens/mar2007/2007-03-06-02.asp

Capriccioso, Rob. 2011. "Indigenous Oil Sands Protest Leads to White House Arrests." Indian Country Today Media Network, September 2. http://indiancountrytoday medianetwork.com/article/indigenous-oil-sands-protest-leads-to-white-house -arrests-51420

Carlton, Jim. 2013. "A Winter without Walrus: Harvesting of Food Staple for Remote Eskimo Villages Plummets; Disaster Declared." *Wall Street Journal*, October 4, A-4.

Cattaneo, Claudia. 2001. "B.C. Natives Blockade Petro-Can Pipeline Site." *National Post*, August 14. http://www.nationalpost.com

Cevallos, Marissa. 2011. "Coal Power? It's Fine—and Have an Inhaler, Says Fake Website." *Los Angeles Times*, May 11. http://articles.latimes.com/2011/may/11/news/la-heb -coal-parody-20110511

"Cheyenne River Sioux Tribe Wins Suit against Homestake." 1999. *Drillbits and Tailings* 4, no. 11 (July 23). http://www.moles.org/ProjectUnderground/drillbits/4_11/2.html

"Chiefs Declare Keystone XL Consultation Meeting Invalid, Walk Out on State Department Officials." 2013. Indian Country Today Media Network, May 17. http://indian countrytodaymedianetwork.com/2013/05/17/chiefs-declare-keystone-xl-consultation -meeting-invalid-walk-out-state-department

Churchill, Ward, and Winona LaDuke. 1986. "Native America: The Political Economy of Radioactive Colonialism." *The Insurgent Sociologist* 13, no. 3 (Spring): 51–84.

Chythlook-Sifsof, Callan J. 2013. "Native Alaska, under Threat." *New York Times*, June 28. http://www.nytimes.com/2013/06/28/opinion/native-culture-under-threat.html

"Clean Coal's Dirty Facts." N.d. http://www.blackmesais.org/clean_coals_dirty_facts.html

"Cleaning Up King Coal? Tribes Appeal Tighter Restrictions on Power Plant Emissions." 2011. Indian Country Today Media Network, May 26. http://indiancountrytoday medianetwork.com/article/cleaning-up-king-coal%3F-tribes-appeal-tighter-restrictions -on-power-plant-emissions-35965

"Cleaning Up Uranium in the Navajo Nation." 2011. Indian Country Today Media Network, May 31. http://indiancountrytodaymedianetwork.com/article/cleaning -up-uranium-in-the-navajo-nation-36290

"Cleanup Planned for Largest Abandoned Uranium Mine on the Navajo Nation." 2011.Environment News Service, September 29. http://www.ens-newswire.com/ ens/sep2011/2011-09-29-092.html

"Climate Change: Arctic Ocean Is Acidifying Fast, Study Shows." 2013. Indian Country Today Media Network, May 15. http://indiancountrytodaymedianetwork.com /2013/05/15/climate-change-arctic-ocean-acidifying-fast-study-shows-149369

"Climate Change Puts Health of Arctic Villagers on Thin Ice." 2011. Indian Country Today Media Network, March 7. http://indiancountrytodaymedianetwork.com/article/ climate-change-puts-health-of-arctic-villagers-on-thin-ice-21391

Cohen, Faye G. 1986. *Treaties on Trial: The Continuing Controversy over Northwest Fishing Rights*. Seattle: University of Washington Press.

Colomeda, Lori A. 1998. "Indigenous Health." Speech. Brisbane, Australia, September 9. http://www.ldb.org/vl/ai/lori_b98.htm

"Colville Tribes Savor Teck Resources Court Admission That It Polluted Columbia River." 2012. Indian Country Today Media Network, October 9. http://indiancountry todaymedianetwork.com/article/colville-tribes-savor-teck-resources-court-admission -that-it-polluted-columbia-river-138899

"Company That Designed Burst B.C. Tailings Pond Was Hired by Pebble Mine in Bristol Bay." 2014. Indian Country Today Media Network, August 8. http://indiancountry

todaymedianetwork.com/2014/08/08/company-designed-burst-bc-tailings-pond-was-hired-pebble-mine-bristol-bay-156315

Cone, Marla. 1996. "Human Immune Systems May Be Pollution Victims." *Los Angeles Times*, May 13, A-1.

Cone, Marla. 2005. *Silent Snow: The Slow Poisoning of the Arctic*. New York: Grove Press.

"Conférence de presse de M. Bernard Landry, premier ministre du Québec, et de M. Ted Moses, grand chef du Grand Conseil des cris Le mardi 23 octobre 2001." 2001. Government of Quebec. October 23. http://www.assnat.qc.ca/fra/conf-presse/2001/011023BL.htm

"Contaminants Have Variety of Effects on Arctic Baby IQs." 2014. Environmental Health News, February 7. http://www.environmentalhealthnews.org/ehs/newscience/2014/Feb/arctic-baby-iqs

Coon-Come, Matthew. 2001. "Remarks of the National Chief Matthew Coon-Come." Peoples Summit of the Americas Environment Forum, April 18. http://www.afn.ca/Press%20Realeses%20%20speeches/april_18.htm

Copland, L., D. R. Mueller, and L. Weir. 2007. "Rapid Loss of the Avies Ice Shelf, Ellesmere Island, Canada." *Geophysical Research Letters* 34: L21501.

Cortes, Amber. N.d. "Treaty Rights and Totem Poles: How One Tribe Is Carving Out a Resistance to Coal." *Grist: A Beacon in the Smog*. http://exp.grist.org/lummi

Cozzetto, K., K. Chief, K. Dittmer, M. Brubaker, R. Gough, K. Souza, F. Ettawageshik, S. Wotkyns, S. Opitz-Stapleton, S. Duren, and P. Chavan. 2013. "Climate Change Impacts on the Water Resources of American Indians and Alaska Natives in the U.S." *Climatic Change* 120: 569–84.

"Cree Claim Victory in Hydro Hearings." 2000. Canadian Broadcasting Corporation, December 1.http://www.northstar.sierraclub.org/HYDRO_Cree_Claim%20_%20Victory.htm

"Cree Indians Win Battle in Their Struggle to Stop Canadian Hydropower Plant." 1991. *Omaha World-Herald*, September 12, 4.

"Cross Lake Takes Hydro Battle to U.S." 2001. First Perspectives: News of Indigenous Peoples of Canada, May 29. http://collection.nlc-bnc.ca/100/201/300/first_perspective/2001/05-29/story7.html

Csordas, T. J. 1999. "Ritual Healing and the Politics of Identity in Contemporary Navajo Society." *American Ethnologist* 26: 3–23.

Dadigan, Marc. 2013. "Umatilla Tribe Battles Mega-Loads Headed for Alberta Oil Sands." Indian Country Today Media Network, December 1. http://indiancountrytodaymedianetwork.com/2013/12/11/umatilla-tribe-battles-mega-loads-headed-alberta-oil-sands-152657

Dakotah. 2013. "Taȟča Wakutėpi: Where They Killed Deer." Indian Country Today Media Network, January 19. http://indiancountrytodaymedianetwork.com/2013/01/19/ta%C4%A5%C4%8B-wakut%C4%97pi-where-they-killed-deer-147081

"Dam Breach at Mount Polley Mine in British Columbia." 2014. NASA Earth Observatory, August 17.http://earthobservatory.nasa.gov/IOTD/view.php?id=84202&src=eoa-iotd

Davis, Julie Hirschfeld. 2015. "Obama Takes Climate Message to Alaska, Where Change Is Rapid." *New York Times*, September 5. http://www.nytimes.com/2015/09/03/us/politics/obama-takes-climate-message-to-alaska-where-change-is-rapid-in-alaska.html

Davis, Neil. 2001. *Permafrost: A Guide to Frozen Ground in Transition*. Fairbanks: University of Alaska Press.

"'Defend Our Climate' Rallies Draw Thousands across Canada." 2013. Environment News Service, November 18. http://ens-newswire.com/2013/11/18/defend-our-climate-rallies-draw-thousands-across-canada

Dewailly, E., P. Ayotte, S. Bruneau, S. Gingras, M. Belles-Isles, and R. Roy. 2000. "Susceptibility to Infections and Immune Status in Inuit Infants Exposed to Organochlorines." *Environment Health Perspectives* 108: 205–11.

Dewailly, E., S. Bruneau, C. Laliberte, M. Belles-Iles, J. P. Weber, and R. Roy. 1993. "Breast Milk Contamination by PCB and PCDD/Fs in Arctic Quebec: Preliminary Results on the Immune Status of Inuit Infants." *Organohalogen Compounds* 13: 403–6.

Dewailly, E., J. J. Ryan, C. Laliberte, S. Bruneau, J. P. Weber, S. Gringras, and G. Carrier. 1994. "Exposure of Remote Maritime Populations to Coplanar PCBs." *Environmental Health Perspectives* 102, suppl. 1: 205–9.

Diaz-Romo, Patricia, and Samuel Salinas-Alvarez. N.d. "Migrant Workers and Pesticides. A Poisoned Culture: The Case of the Indigenous Huicholes Farm Workers." *Abya Yala News: The Journal of the South and Meso-American Rights Center*. http://saiic.nativeweb.org/ayn/huichol.html

Dobbyn, Paula. 2000. "Contaminated Game Has Natives Worried." *Anchorage Daily News*, May 2. http://www.adn.com http://www.ienearth.org/food_toxic.html

Dougherty, Kevin. 2001. "Deal Is No Blank Cheque, Crees Say." *Montreal Gazette*, October 24. http://www.canada.com/montreal/montrealgazette/story.asp?id={4FCA2E44-DDE4-4675-88C5-DEFD94486151

Doyle, John T., Margaret Hiza Redsteer, and Margaret J. Eggers. 2013. "Exploring Effects of Climate Change on Northern Plains American Indian Health." *Climatic Change* 120: 643–55.

Duffy, Andrew. 2000a. "Global Warming Causing Arctic Town to Sink, Says Inuit Leader 'Warning Signal.'" *Montreal Gazette*, April 18. http://www.climateark.org/articles/2000/2nd/glwatosi.htm

Duffy, Andrew. 2000b. "Global Warming Why Arctic Town Is Sinking: Permafrost Is Melting under Sachs: Inuit Leader." *Montreal Gazette*, April 18, A-13.

Dumont, C. 1995. *Proceedings of 1995 Canadian Mercury Network Workshop. Mercury and Health: The James Bay Cree*. Montreal: Cree Board of Health and Social Services. http://www.cciw.ca/eman-temp/reports/publications/mercury95/part4.html

Dumont, Charles, Manon Girard, François Bellavance, and Francine Noël. 1998. "Mercury Levels in the Cree Population of James Bay, Quebec, from 1988 to 1993/94." *Canadian Medical Association Journal* 158 (June 2): 1439–45. http://www.cma.ca/cmaj/vol-158/issue-11/1439.htm

Ebinghaus, Ralf, Hans H. Kock, Christian Temme, Jürgen W. Einax, Astrid G. Löwe, Andreas Richter, John P. Burrows, and William H. Schroeder. 2002. "Antarctic Springtime Depletion of Atmospheric Mercury." *Environmental Science and Technology* 36, no. 6 (March 15): 1238–44.

Egan, D'Arcy. 2001. "Dam the Ecology If Quebec Goes Full-Speed Ahead." *Plain Dealer*, July 29, D-15.

Eichstaedt, Peter. 1994. *If You Poison Us: Uranium and American Indians*. Santa Fe, NM: Red Crane Books.

Eilperin, Juliet. 2012. "Alaskan Arctic Villages Hit Hard by Climate Change." *Washington Post*, August 5. http://www.washingtonpost.com/national/health-science/alaskan-arctic-villages-hit-hard-by-climate-change/2012/08/05/e9dbd4a6-d5b0-11e1-a9e3-c5249ea531ca_print.html

"Elsipogtog Chief Issues Eviction Notice to Texas-Based Frackers. Band Council Resolution to Reclaim All Unoccupied Crown Land." 2013. Halifax Media Co-op, October 4. http://halifax.mediacoop.ca/story/elsipogtog-chief-issues-eviction-notice-texas-base/1909

"Emissions from Oil Sands Mining." 2012. NASA Earth Observatory, March 2. http://earthobservatory.nasa.gov/IOTD/view.php?id=77283&src=eoa-iotd

Emmons, Della Gould. 1965. *Leschi of the Nisquallies*. Minneapolis: T. S. Denison.

"Energy Revenues to American Indians Top $1 Billion for First Time." 2014. Indian Country Today Media Network, December 16. http://indiancountrytodaymedianetwork.com/2014/12/16/energy-revenues-american-indians-top-1-billion-first-time-158312

"EPA Says Wyoming Fracking Results Are Consistent." 2012. *Wall Street Journal*, September 26. http://online.wsj.com/article/SB10000872396390443328404578020923049282436.html

Erlanger, Steven. 2013. "As Drilling Practice Takes Off in U.S., Europe Proves Hesitant." *New York Times*, October 9. http://www.nytimes.com/2013/10/10/world/europe/as-drilling-practice-takes-off-in-us-europe-proves-hesitant.htm

"'Every Tree and Plant Died': Massive Toxic Spill Guts Alberta." 2013. Indian Country Today Media Network, August 26. http://indiancountrytodaymedianetwork.com/2013/08/26/toxic-wastewater-spill-alberta-kills-dene-tha-landscape-150968

Fahys, Judy. 2011. "Uranium Cleanup Under Way on Navajo Land." *Salt Lake Tribune*, May 25.http://www.sltrib.com/sltrib/politics/51833502-90/begay-cleanup-epa-helen.html.csp

"Feds Reach Out to Natives on Climate Change at Tribal Nations Conference." 2013. Indian Country Media Network, November 17. http://indiancountrytodaymedianetwork.com/2013/11/17/feds-reach-out-natives-climate-change-tribal-nations-conference-152290

Fernandez, Lydia. 1998. "Gold Mine Threatens Quechan Sites." *Native Americas* 15, no. 2 (Summer): 6.

"Figure This One!" 1998. *Four Corners Clamor* 1, no. 6 (Summer/Fall). http://www.ausbcomp.com/redman/clamor6.htm#newdeal

"Fill Gaps in Keystone XL Draft Environment Report or Reject Pipeline, NCAI Tells Obama Administration." 2013. Indian Country Today Media Network, May 2. http://indiancountrytodaymedianetwork.com/2013/05/02/fill-gaps-keystone-xl-draft-environment-report-or-reject-pipeline-ncai-tells-obama.html

"First Stewards Resolution Asks U.S. Congress to Formally Recognize Ecological Knowledge of Coastal Indigenous People." 2012. Indian Country Today Media Network, September 30. http://indiancountrytodaymedianetwork.com/article/first-stewards-resolution-asks-u.s.-congress-to-formally-recognize-ecological-knowledge-of-coastal-indigenous-people-136802

"$4.3 Million Settlement over Reid-Gardner Coal Plant for Moapa Paiute." 2015. Indian Country Today Media Network, August 6. http://indiancountrytodaymedianetwork.com/2015/08/06/43-million-settlement-over-reid-gardner-coal-plant-moapa-paiute-161299

"'Freedom Train' Protesting Northern Gateway Pipeline Crossing Canada." 2012. Indian Country Today Media Network, April 30. http://indiancountrytodaymedianetwork.com/most-popular?page=2677

Frey, Darcy. 2002. "George Divoky's Planet." *New York Times Sunday Magazine*, January 6, 26–30.

Frosch, Dan. 2013. "Amid Pipeline Debate, Two Costly Cleanups Forever Change Towns." *New York Times*, August 10. http://www.nytimes.com/2013/08/11/us/amid-pipeline-debate-two-costly-cleanups-forever-change-towns.html.

Frosch, Dan. 2014. "Amid Toxic Waste, a Navajo Village Could Lose Its Land." *New York Times*, February 20, A-9, A-12. http://www.nytimes.com/2014/02/20/us/nestled-amid-toxic-waste-a-navajo-village-faces-losing-its-land-forever.html

"Full Text of the Great Plains Tribal Chairmen's Association Statement against the Keystone XL Pipeline." 2013. Indian Country Today Media Network, May 17. http://indiancountrytodaymedianetwork.com/2013/05/17/full-text-great-plains-tribal-chairmens-association-statement-against-keystone-xl

Galchen, Rivka. 2015. "Weather Underground." *The New Yorker*, April 13, 34–40.

Gallo, Mia B., Lawrence M. Schell, Anthony P. DeCaprio, and Agnes Jacobs. 2001. "Levels of Persistent Organic Pollutant and Their Predictors among Young Adults." *Chemosphere* 83, no. 10 (May): 1374–82.

Gedicks, Al. 1993. *The New Resource Wars: Native and Environmental Struggles against Multinational Corporations*. Boston: South End Press.

George, Jane. 2000. "Sierra Club Focuses on Arctic Global Warming." *Nunatsiaq News*, January 31. http://www.nunatsiaq.com/archives/nunavut000131/nvt20121_04.html

Gibson, Leah. 2013. "Oregon Tribes Await Superfund Attention for Portland Harbor Site." Indian Country Today Media Network, February 1.http://indiancountrytodaymedianetwork.com/2013/02/01/oregon-tribes-await-superfund-attention-portland-harbor-site-147374

Gilday, Cindy Kenny. N.d. "A Village of Widows." *Arctic Circle*. http://arcticcircle.uconn.edu/SEEJ/Mining/gilday.html

Gillis, Justin. 2012. "Sea Ice in Arctic Measured at Record Low." *New York Times*, August 27. http://www.nytimes.com/2012/08/28/science/earth/sea-ice-in-arctic-measured-at-record-low.html

Gillis, Justin, and Henry Fountain. 2014. "Trying to Reclaim Leadership on Climate Change." *New York Times*, June 2, A-12.

Giuliano, Jackie Alan. 2001. "Killing Tomorrow for a Few Megawatts Today." Environment News Service, March. http://ens.lycos.com/ens/mar2001/2001L-03-30g.html

"Global Warming Worries Native Americans." 1998. *Deseret News*, November 27. http://www.desnews.com/cit/071en14f.htm

"Goldman Prize Winner Caroline Cannon." 2012. National Public Radio. "Living on Earth." April 20. http://www.loe.org/shows/segments.html?programID=12-P13-00016&segmentID=5

Goldtooth, Tom. 2001. "Indigenous Environmental Network Statement to Nora Helton, Chairwoman, Fort Mojave Indian Tribe, Other Lower Colorado River Indian Tribal Leaders, Tribal Community Members, Elders and Non-Native Groups and Individuals, in Reference to the Ward Valley Victory Gathering to Celebrate the Defeat of a Proposed Nuclear Waste Dump, Ward Valley, California." February 16. http://www.ienearth.org/ward_valley2.html

Government Accountability Office. 2012. "GAO Report: $3.5 Billion Worth of Minerals Produced on Indian Lands in FY 2011." Indian Country Today Media Network, December 13. http://indiancountrytodaymedianetwork.com/article/gao-report-35-billion-worth-minerals-produced-indian-lands-fy-2011-146271

Green, Sara Jean. N.d. "Fighting a GIANT." *Windspeaker*. http://www.ammsa.com/classroom/CLASS3Lubicon.html

Grinde, Donald A., Jr., and Bruce E. Johansen. 1995. *Ecocide of Native America: Environmental Destruction of Indian Lands and Peoples*. Santa Fe, NM: Clear Light.

Grossman, Zoltán, and Alan Parker. 2012. *Asserting Native Resilience: Pacific Rim Indigenous Nations Face the Climate Crisis*. Corvallis: Oregon State University Press.

Guerette, Deb. 2001. "No Clear-Cut Answer: Timber Rights Allocation on Lubicon Land a Worrisome Development." *Grande Prairie Daily Herald-Tribune*, March 5, n.p.

Guillette, Elizabeth A., Maria Mercedes Meza, Maria Guadalupe Aquilar, Alma Delia Soto, and Idalia Enedina Garcia. 1998. "An Anthropological Approach to the Evaluation of Preschool Children Exposed to Pesticides in Mexico." *Environmental Health Perspectives* 106, no. 6 (June). http://www.anarac.com/elizabeth_guillette.htm

Hand, Eric. 2014. "Injection Wells Blamed in Oklahoma Earthquakes." *Science* 345 (July 4): 13–14.

Hansen, James. 2012. "Game Over for the Climate." *New York Times*, May 9. http://www.nytimes.com/2012/05/10/opinion/game-over-for-the-climate.html

Hansen, James, Reto Ruedy, Makiko Sato, and Ken Lo. 2010. "Global Temperature and Europe's Frigid Air." Draft paper, December 10.

Hansen, Terri. 2011. "Akwesasne Mohawk Youth Are Still at Risk of Industrial Pollutants." Indian Country Today Media Network, June 20. http://indiancountrytodaymedianetwork.com/2011/06/20/akwesasne-mohawk-youth-are-still-risk-industrial-pollutants-39158

Hansen, Terri. 2012. "Northwest Tribes Fight for Treaty Rights in Face of Coal-Transport Plan." Indian Country Today Media Network, August 15. http://indiancountrytodaymedianetwork.com/article/northwest-tribes-fight-for-treaty-rights-in-face-of-coal-transport-plan-129453

Hansen, Terri. 2013. "Major Environmental Disasters in Indian Country." Indian Country Today Media Network, October 8. http://indiancountrytodaymedianetwork.com/2013/10/08/7-major-industrial-environmental-disasters-indian-country-151661

Hansen, Terri. 2014a. "Climate Disruptions Hitting More and More Tribal Nations." Indian Country Today Media Network, May 7. http://indiancountrytodaymedianetwork.com/2014/05/07/climate-disruptions-hitting-more-and-more-tribal-nations-154747

Hansen, Terri. 2014b. "Coast Salish Unite against Tripling Capacity of Kinder Morgan Tar Sands Pipeline." Indian Country Today Media Network, October 27. http://indiancountrytodaymedianetwork.com/2014/10/27/coast-salish-unite-against-tripling-capacity-kinder-morgan-tar-sands-pipeline-157543

Hansen, Terri. 2014c. "532 Superfund Sites in Indian Country." Indian Country Today Media Network, June 17. http://indiancountrytodaymedianetwork.com/2014/06/17/532-superfund-sites-indian-country-155316

Harney, Corbin. 1995. *The Way It Is: One Water, One Air, One Earth*. Nevada City, CA: Blue Dolphin Publishing.

"Havasupai Fight to Save Grand Canyon from Uranium Mining." 1992. July 21.http://www.ratical.org/ratville/native/havasuEFN.txt

Healy, Jack. "Tapping into the Land, and Dividing Its People." *New York Times*, August 15. http://www.nytimes.com/2012/08/16/us/montana-tribe-divided-on-tapping-oil-rich-land.html

"Heavy Metal Levels in Reindeer, Caribou, and Plants of the Seward Peninsula." 2000. Current Research Programs, Reindeer Research Program, University of Alaska Fairbanks, April. http://reindeer.salrm.alaska.edu/research.htm

Herbert, H. Josef. 2000. "Inuit Say They Are Witness to Global Warming in the Arctic." *Milwaukee Journal-Sentinel*, November 19. http://www.jsonline.com/alive/news/nov00/warming20111900.asp

Hernandez, Ruben. 2012. "An Energy Revolution on the Navajo Reservation." New America Media. December 10. http://newamericamedia.org/2012/12/an-energy-revolution-on-the-navajo-reservation.php

Hill, Miriam. 2001. "Iqaluit's Waste Woes Won't Go Away; City Sets Up Bins Where Residents Can Dump Plastics, Metal." *Nunatsiag News*, July 27, 5.

Hodge, Damon. 2001. "Dumpy Meeting: Final Vegas Hearing on Yucca Mountain Plagued by the D.O.E." *Las Vegas Weekly*, September 11. http://www.lasvegasweekly.com/2001_2/09_13/news_upfront7.html

Homer-Dixon, Thomas. 2013. "The Tar Sands Disaster." *New York Times*, March 31. http://www.nytimes.com/2013/04/01/opinion/the-tar-sands-disaster.html

Hooks, Gregory, and Chad L. Smith. 2004. "The Treadmill of Destruction: National Sacrifice Areas and Native Americans." *American Sociological Review* 69, no. 4: 558–75.

"Horrific Toxic Spill in B.C. Called Another Exxon Valdez." 2014. Indian Country Today Media Network, August 7. http://www.nativenewstoday.com/2014/08/07/horrific-toxic-spill-in-b-c-called-another-exxon-valdez

"Hotspots: California." 2001. *Drillbits and Tailings* 6, no. 9 (November 30). http://www.moles.org

"Hotspots: Idaho." 2001. *Drillbits and Tailings* 6, no. 7 (October 31): 3.

Howe, Miles. 2013. "RCMP Bring 60 Drawn Guns, Dogs, Assault Rifles, to Serve Injunction on the Wrong Road." Halifax Media Co-op, October 18. http://halifax.mediacoop.ca/story/rcmp-bring-60-drawn-guns-dogs-assault-rifles-serve/19358

"Hudson Bay Ice-Free by 2050, Scientists Say." 2001. Canadian Broadcasting Corporation, March 14. http://north.cbc.ca/cgi-bin/templates/view.cgi?/news/2001/03/14/14hudsonice

Huff, Andrew. 2000. "Gold Mining Threatens Communities." The Progressive Media Project, July 11. http://www.progressive.org/mpdvah00.htm

Hughes, Pam. 2013a. "Bad River Band Establishes Legal Defense Fund to Stop Proposed Iron-Ore Mine." Indian Country Today Media Network, March 19. http://indiancountrytodaymedianetwork.com/2013/03/19/bad-river-band-establishes-legal-defense-fund-stop-proposed-iron-ore-mine-148251

Hughes, Pam. 2013b. "Wisconsin Tribes: What Part of 'No Mining!' Don't You Understand?" Indian Country Today Media Network, July 7. http://indiancountrytodaymedianetwork.com/2013/07/07/wisconsin-tribes-what-part-no-mining-dont-you-understand-150309

Hulen, David. 1994. "Hunt Is On for Pollutant Traces in Bering Sea; Alaska Villagers, Scientists Wonder If Toxic Substances Are Endangering Animals and People Who Eat Them." *Los Angeles Times*, August 15, A-5.

Human, Katy. 2004. "Disappearing Arctic Ice Chills Scientists; A University of Colorado Expert on Ice Worries That the Massive Melting Will Trigger Dramatic Changes in the World's Weather." *Denver Post*, October 5, B-2.

"Hydroelectric Production Is Anything but Cheap." 2001. *Virtual Circle: First Nations' Chronicles*, March 21.http://www.vcircle.com/journal/archive/010.shtml

"Ice a Scarce Commodity on Arctic Rinks: Global Warming Blamed for Shortened Hockey Season." 2003. *Financial Post*, January 7, A-3.

"Indigenous Alaskans Shed Light on Climate Change." 2011. Indian Country Today Media Network, September 14. http://indiancountrytodaymedianetwork.com/article/indigenous-alaskans-shed-light-on-climate-change-53921

"In Historic Visit, Obama Vows to Help." 2014. *Omaha World-Herald*, June 14, A-1, A-3.

"Iñupiat Hunting Community Divided over Shell Drilling in Chukchi Sea off Alaska." 2012. Indian Country Today Media Network, October 4. http://indiancountrytodaymedianetwork.com/article/i%C3%B1upiat-hunting-community-divided-over-shell-drilling-in-chukchi-sea-off-alaska-137125

"Iqaluit Drunks Keep Cops' Hands Full." 2001. CBC North/News, March 16. http://north.cbc.ca/cgi-bin/templates/view.cgi?/news/2001/03/16/16iqalcohol

Ivins, Molly K. 1978. "Doctor Charges U.S. Ignored Uranium Spill." *Hutchinson (Kans.) News*, July 27, n.p.

Jacobson, Joseph L., and Sandra W. Jacobson. 1996. "Intellectual Impairment in Children Exposed to Polychlorinated Biphenyls in Utero." *New England Journal of Medicine* 335, no. 11: 783–89.

Jaimet, Kate. 2001. "Canada to Fight Bush over Arctic Oil Drilling: Environment Minister 'Utterly Opposed' to Exploitation of Wildlife Refuge." *Ottawa Citizen*, January 4, A-1.

James, Jewell. 2014. "2014 Totem Pole Journey Honors Tribes' Stewardship of Land, Water." *Bellingham Herald*, August 11. http://www.bellinghamherald.com/2014/08/11/3792147_2014-totem-pole-journey-honors.html

Jans, Nick. 2008. "Living with Oil: The Real Price." In *The Last Polar Bear: Facing the Truth of a Warming World*, edited by Steven Kazlowski, 147–61. Seattle: Braided River.

Jessepe, Lorraine. 2012. "Alaskan Native Communities Facing Climate-Induced Relocation." Indian Country Today Media Network, June 21. http://indiancountrytodaymedianetwork.com/article/alaskan-native-communities-facing-climate-induced-relocation-119615

Johansen, Bruce E. 1993. *Life and Death in Mohawk Country*. Golden, CO: North American Press/Fulcrum.

Johansen, Bruce E. 1995. "Victims of Progress: Navajo Miners" [Review of Eichstaedt, *If You Poison Us*]. *Native Americas* 12, no. 3 (Fall): 60–61.

Johansen, Bruce E. 1997. "The High Cost of Uranium in Navajoland." *Akwesasne Notes New Series* 2, no. 2 (Spring): 10–12. http://ratical.com/radiation/UraniumInNavLand.html

Johansen, Bruce E. 1998a. "Exxon vs. Chippewas in Wisconsin." *Native Americas* 15, no. 1 (Spring): 6.

Johansen, Bruce. 1998b. "Ward Valley: A 'Win' for Native Elders." *Native Americas* 15, no. 2 (Summer): 7–8.

Johansen, Bruce E. 2000. "Pristine No More: The Arctic, Where Mother's Milk Is Toxic." *The Progressive*, December, 27–29.

Johansen, Bruce E. 2001. "Arctic Heat Wave." *The Progressive*, October, 18–20.

Johansen, Bruce, and Roberto Maestas. 1979. *Wasi'chu: The Continuing Indian Wars*. New York: Monthly Review Press.

Johnson, Kirk. 2013a. "Fight over Energy Finds a New Front in a Corner of Idaho." *New York Times*, September 25. http://www.nytimes.com/2013/09/26/us/fight-over-energy-finds-a-new-front-in-a-corner-of-idaho.html

Johnson, Kirk. 2013b. "With Proposed Rail Expansion, Northwest Confronts Its Clean Image." *New York Times*, August 13. http://www.nytimes.com/2013/08/20/us/with-proposed-rail-expansion-northwest-confronts-its-clean-image.html

Joling, Dan. 2007a. "Thousands of Pacific Walruses Die; Global Warming Blamed." Associated Press, December 14.

Joling, Dan. 2007b. "Walruses Abandon Ice for Alaska Shore." *Washington Post*, October 4. http://www.washingtonpost.com/wp-dyn/content/article/2007/10/04/AR2007100402299_pf.html

"Just Say No Mega-Load! Columbia River Inter-Tribal Fish Commission Joins Protest." 2013. Indian Country Today Media Network, September 9. http://indiancountrytodaymedianetwork.com/2013/09/09/columbia-river-inter-tribal-fish-commission-joins-anti-mega-load-chorus-resolution-151203

Kaufman, Dan. 2014. "The Fight for Wisconsin's Soul." *New York Times*, March 29. http://www.nytimes.com/2014/03/30/opinion/sunday/the-fight-for-wisconsins-soul.html

Keranen, K. M., M. Weingarten, G. A. Abers, B. A. Berkins, and S. Ge. 2014. "Sharp Increase in Central Oklahoma Seismicity since 2008 Induced by Massive Wastewater Injection." *Science* 345 (July 25): 444–48.

Kerr, Richard A. 2012. "Ice-Free Arctic Sea May Be Years, Not Decades, Away." *Science* 337 (September 28): 1591.

Kizzia, Tom. 1998. "Seal Hunters Await Late Ice." *Anchorage Daily News*, November 28. http://www.adn.com/stories/T98112872.html

Klein, Naomi. 2014. *This Changes Everything: Capitalism and the Climate*. New York: Simon & Schuster.

Knight, Danielle. 2000. "Inuit Tell Negotiators of Climate Change Impact." Interpress Service, November 16.

Knight, Danielle. 2013. "Native Americans Denounce Toxic Legacy." TWN: Third World Network, June 14.http://www.twnside.org.sg/title/legacy-cn.htm

Koch, Wendy. 2014. "Would Keystone Pipeline Unload 'Carbon Bomb' or Job Boom?" *USA Today*, March 10. http://www.usatoday.com/story/news/nation/2014/03/01/keystonexls-myths-debunked/5651099

Kolbert, Elizabeth. 2007a. "Testing the Climate." *The New Yorker*, December 24 and 31, 43–44.

Kolbert, Elizabeth. 2007b. "Unconventional Crude: Canada's Synthetic-Fuels Boom." *The New Yorker*, November 12, 46–51.

Kort, Eric A., Christian Frankenberg, Keeley R. Costigan, Rodica Lindenmaier, Manvendra K. Dubey, and Debra Wunch. 2014. "Four Corners: The Largest U.S. Methane Anomaly Viewed from Space." *Geophysical Research Letters* 41, no. 19 (October 16): 6898–903.

Krajick, Kevin. 2001. "Arctic Life, on Thin Ice." *Science* 291 (January 19): 424–25.

Krauss, Clifford. 2013. "Looking for a Way around Keystone XL, Canadian Oil Hits the Rails." *New York Times*, October 31. http://www.nytimes.com/2013/10/31/business/energy-environment/looking-for-a-way-around-keystone-xl-canadian-oil-hits-the-rails.html

Krauss, Clifford. 2014. "U.S. Oil Prices Fall below $80 a Barrel." *New York Times*, November 4, B-3.

Krauss, Clifford, and Ian Austen. 2014. "Rocky Road for Canadian Oil." *New York Times*, May 13, B-1, B-4.

Krauss, Clifford, and Jad Mouawad. 2014. "Accidents Surge as Oil Industry Takes the Train." *New York Times*, January 26, A-1.

"Lac-Mégantic Rail Tragedy Resonates at Quinault Nation." 2013. Indian Country Today Media Network, July 29. http://indiancountrytodaymedianetwork.com/2013/07/

29/lac-megantic-rail-tragedy-resonates-quinault-nation-victims-are-memorialized-150626

LaDuke, Winona. 1992. "Indigenous Environmental Perspectives: A North American Primer." *Native Americas* 9, no. 2 (Summer): 52–71.

LaDuke, Winona. 1993. "Tribal Coalition Dams Hydro-Quebec Project." *Indian Country Today*, July 21, A-3.

LaDuke, Winona. 1994. "Breast Milk, P.C.B.s, and Motherhood: An Interview with Katsi Cook, Mohawk." *Cultural Survival Quarterly* 17, no. 4 (Winter): 43–45.

LaDuke, Winona. 2001. "Insider Essays: Our Responsibility." Electnet/Newswire, October 2.http://www.electnet.org/dsp_essay.cfm

LaDuke, Winona. 2014a. "Bakken Gas Flares Away, as Nationwide Propane Shortage Kills with Hypothermia." Indian Country Today Media Network, October 9. http://indiancountrytodaymedianetwork.com/2014/10/09/bakken-gas-flares-away-nationwide-propane-shortage-kills-hypothermia-157249

LaDuke, Winona. 2014b. "'Unspeakable Poverty of Loss': Intergenerational Trauma and the Bakken Oil Fields." Indian Country Today Media Network, October 8. http://indiancountrytodaymedianetwork.com/2014/10/08/unspeakable-poverty-loss-intergenerational-trauma-and-bakken-oil-fields-157243

LaDuke, Winona, and Sean Aaron Cruz. 2013. *The Militarization of Indian Country*. East Lansing: Michigan State University Press.

Lamb, David Michael. N.d. "Toxins in a Fragile Frontier." Transcript. Canadian Broadcasting Corporation News.http://cac.ca/news/indepth/north

Landry, Alysa. 2014. "Is the Navajo Mine a Viable Option? Growing Opposition Doesn't Think So." Indian Country Today Media Network, February 6. http://indiancountrytodaymedianetwork.com/2014/02/06/navajo-mine-viable-option-growing-opposition-doesnt-think-so-153407

Lazenby, Henry. 2014. "Taseko Stocks Tumble as Fed Rejects New Prosperity, BC—Again." *Mining Weekly*, February 27. http://www.miningweekly.com/article/taseko-stocks-down-as-fed-rejects-new-prosperity-bc-again-2014-02-27

Lee, Jaeah. 2011. "In Arctic, Warmer Climate Means More Beavers Defecating and Disease." *Mother Jones*, March 4.

Lee, Tanya. 2013. "Important Cultural, Religious and Historical Resources Threatened by Drilling." Indian Country Today Media Network, April 8. http://indiancountrytodaymedianetwork.com/2013/04/08/important-cultural-religious-and-historical-resources-threatened-drilling-148622

Lee, Tanya H. 2014. "H-Bomb Guinea Pigs! Natives Suffering Decades after New Mexico Tests." Indian Country Today Media Network, March 5. http://indiancountrytodaymedianetwork.com/2014/03/05/guinea-pigs-indigenous-people-suffering-decades-after-new-mexico-h-bomb-testing-153856?page=0%2C3

LeMay, Connie. 2013. "No Longer Dead in the Water: Last Year's Ailing Wild Rice Crop Bounces Back." Indian Country Today Media Network, September 23. http://www.Indiancountrytodaymedianetwork.com/2013/09/23/no-longer-dead-water-last-years-ailing-wild-rice-crop-bounces-back-151402

"Leo DiCaprio Tours Tar Sands, Joins Natives in ALS Ice Bucket Challenge." 2014. Indian Country Today Media Network, August 26. http://indiancountrytodaymedianetwork.com/2014/08/26/leo-dicaprio-tours-tar-sands-joins-natives-als-ice-bucket-challenge-156602

Leslie, Jacques. 2014. "Is Canada Tarring Itself?" *New York Times*, March 31, A-19.

Lickers, Henry. 1995. "Guest Essay: The Message Returns." *Akwesasne Notes New Series* 1, no. 1 (April–June): 10–11.

Lindberg, Steve E., Steve Brooks, C.-J. Lin, Karen J. Scott, Matthew S. Landis, Robert K. Stevens, Mike Goodsite, and Andreas Richter. 2002. "Dynamic Oxidation of Gaseous Mercury in the Arctic Troposphere at Polar Sunrise." *Environmental Science and Technology* 36, no. 6 (March 15): 1245–56.

Lizza, Ryan. 2013. "The President and the Pipeline." *The New Yorker*, September 16, 38–51.

Ljunggren, David. 2000. "Effects of Global Warming Clear in Canada Arctic." Environmental News Network, April 20. http://www.enn.com/enn-subsciber-news-archive/2000/04/04202000/reu_arctwarm_12170.asp

"Lower Brule Sioux Reject Keystone XL, Evict TransCanada." 2015. Indian Country Today Media Network, May 1. http://indiancountrytodaymedianetwork.com/2015/05/01/lower-brule-sioux-reject-keystone-xl-evict-transcanada-160214

"Lummi Nation Officially Opposes Coal Export Terminal in Letter to Army Corps of Engineers." 2012. Indian Country Today Media Network, August 2. http://indiancountrytodaymedianetwork.com/2013/08/02/lummi-nation-officially-opposes-coal-export-terminal-letter-army-corps-engineers-150718

Luoma, Jon R. 1999. "System Failure: The Chemical Revolution Has Ushered In a World of Changes. Many of Them, It's Becoming Clear, Are in Our Bodies." *Mother Jones*, July/August. http://www.motherjones.com/mother_jones/JA99/endocrine/html

Lustgarten, Abraham. 2015. "Inside the Power Plant Fueling America's Drought." ProPublica in Indian Country Today Media Network, July 3. http://indiancountrytodaymedianetwork.com/2015/07/03/inside-power-plant-fueling-americas-drought-160928

Lynn, Kathy, John Daigle, Jennie Hoffman, Frank Lake, Natalie Michelle, Darren Ranco, Carson Viles, Garrit Voggesser, and Paul Williams. 2013. "The Impacts of Climate Change on Tribal Traditional Foods." *Climatic Change* 120: 545–56.

Macafee, Michelle. 2001. "Cree Grand Chief Defends Hydro Deal with Quebec as Negotiations Wind Down." *Ottawa Citizen*, December 18. http://www.canada.com/news/story.asp?id={565BD3CB-8DAB-4CD2-878F-EE2F7807666B}

MacMillan, Leslie. 2012. "Uranium Mines Dot Navajo Land, Neglected and Still Perilous." *New York Times*, March 31. http://www.nytimes.com/2012/04/01/us/uranium-mines-dot-navajo-land-neglected-and-still-perilous.html

"Major Victory: Canadian Supreme Court Hands Tsilhqot'in Aboriginal Title." 2014. Indian Country Today Media Network, June 26. https://indiancountrytodaymedianetwork.com/2014/06/26/major-victory-canadian-supreme-court-hands-tsilhqotin-aboriginal-title-155493

Malakoff, David. 2014. "Oil Sands Fight Heats Up in U.S." *Science* 345 (July 25): 368–69.

Mann, Judy. 2000. "A Cautionary Tale about Pesticides." *Washington Post*, June 2, C-9.

Maslanik, J. A., M. C. Serreze, and T. Agnew. 1999. "On the Record Reduction in 1998 Western Arctic Sea Ice Cover." *Geophysical Research Letters* 26, no. 13: 1905,1999.

Mathews-Amos, Amy, and Ewann A. Berntson. 1999. "Turning Up the Heat: How Global Warming Threatens Life in the Sea." World Wildlife Fund and Marine Conservation Biology Institute. http://www.worldwildlife.org/news/pubs/wwf_ocean.htm

Matteo, Enzo di. 2002. "Damned Deal: Cree Leaders Call Hydro Pact Signed in Secret a Monstrous Sellout." *Now Magazine*, February. http://www.nowtoronto.com/issues/2002-02-14/news_story.php

McGrath, Susan. 2011. "On Thin Ice." *National Geographic*, July, 64–75.

McKeown-Eyssen G. E., and J. Ruedy. 1983. "Methyl-Mercury Exposure in Northern Quebec: Neurologic Findings in Adults." *American Journal of Epidemiology* 118: 461–69.

McLinden, C. A., V. Fioletov, K. F. Boersma, N. Krotkov, C. E. Sioris, J. P. Veefkind, and K. Yang. 2012. "Air Quality over the Canadian Oil Sands: A First Assessment Using Satellite Observations." *Geophysical Research Letters* 39: L04804.

McNeel, Jack. 2012. "Railroad to Disaster? Inland Tribes Fight to Avert Coal-Train Destruction." Indian Country Today Media Network, December 17. http://indiancountrytodaymedianetwork.com/article/railroad-disaster-inland-tribes-fight-avert-coal-train-destruction-146338

McNeel, Jack. 2013. "Watershed Heroes: Colville Confederated Tribes Win Sierra Club Award for Battling British Columbia Smelter." Indian Country Today Media Network, March 6. http://indiancountrytodaymedianetwork.com/2013/03/06/watershed-heroes-colville-confederated-tribes-win-sierra-club-award-battling-british

McNeel, Jack. 2015. "Too Many Broken Pipelines: Fort Peck Reservation Passes Resolution Opposing Keystone XL through Montana." Indian Country Today Media Network, February 24. http://indiancountrytodaymedianetwork.com/2015/02/24/too-many-broken-pipelines-fort-peck-reservation-passes-resolution-opposing-keystone-xl

"Mercury Contaminates Polar Regions." 2002. Environment News Service, March 21. http://ens-news.com/ens/mar2002/2002L-03-21-09.html#anchor1

"Mercury Poisoning Five Decades Later for Grassy Narrows and White Dog First Nation." 2012. Indian Country Today Media Network, June 12. http://indiancountrytodaymedianetwork.com/article/mercury-poisoning-five-decades-later-for-grassy-narrows-and-white-dog-first-nations-117843

"Message in a Blanket: Dakota Artist Stealth-Embroiders 'NOKXL' into Obama Quilt." 2015. Indian Country Today Media Network, May 16. http://indiancountrytodaymedianetwork.com/2015/05/16/message-blanket-dakota-artist-embroiders-nokxl-obama-quilt-160385

"Militarization and Oil in Chiapas, Mexico." *Drillbits and Tailings* 4, no, 16 (October 8). http://www.moles.org/ProjectUnderground/drillbits/4_16/3.html

Millet, Lydia. 2015. "Selling Off the Holy Land." *New York Times*, May 29, A-23, A-27.

Minard, Anne. 2013. "Arizona Tribes and Environmental Groups Oppose HR 287 for Mining in the Tonto National Forest." Indian Country Today Media Network, February 26.http://indiancountrytodaymedianetwork.com/2013/02/26/arizona-tribes-and-environmental-groups-oppose-hr-287-mining-tonto-national-forest-147894

Minard, Anne. 2014. "Methane 'Hot Spot' Seen from Space Hovers over Four Corners." Indian Country Today Media Network, October 28. http://indiancountrytodaymedianetwork.com/2014/10/28/methane-hot-spot-seen-space-hovers-over-four-corners-157560

"Mining Company to Pay Coeur d'Alene, State of Idaho and U.S. Government." 2011. Indian Country Today Media Network, June 16. http://indiancountrytodaymedianetwork.com/article/mining-company-to-pay-coeur-d%E2%80%99alene%2C-state-of-idaho-and-u.s.-government-38602

"Moapa Paiute March 50 Miles in Anti-Coal Protest." 2012. Indian Country Today Media Network, April 27. http://indiancountrytodaymedianetwork.com/article/moapa-paiute-march-50-miles-in-anti-coal-protest-110450

"Moapa Paiute Sue over Coal Plant Contaminants." 2013. Indian Country Today Media Network, August 9. http://indiancountrytodaymedianetwork.com/2013/08/09/moapa-paiute-sue-over-coal-plant-contaminants-150806

"Moapa Paiute to Host Second Solar Project on Its Lands." 2014. Indian Country Today Media Network, May 29. http://indiancountrytodaymedianetwork.com/2014/05/29/moapa-paiute-host-2nd-solar-project-its-lands-155071

Mofina, Rick. 2000. "Study Pinpoints Dioxin Origins: Cancer-Causing Agents in Arctic Aboriginals' Breast Milk Comes from U.S. and Quebec." *Montreal Gazette*, October 4, A-12.

"Mohawk Government Opposes Grasse River Cleanup Plan." 2012. *Watertown Daily Times*, October 2. http://www.watertowndailytimes.com/article/20121002/NEWS09/710029710

"Mohawk Man Arrested for Taking Backhoe to Superfund Site." 2011. Indian Country Today Media Network, August 12. http://indiancountrytodaymedianetwork.com/article/mohawk-man-arrested-for-taking-backhoe-to-superfund-site-46891

"Mohawks Express Gratitude for Akwesasne Health Advocate." 2012. Indian Country Today Media Network, January 6. http://indiancountrytodaymedianetwork.com/article/mohawk-man-fights-to-remove-toxic-hazardous-waste-from-river-104947

"Mohawks Say EPA Alcoa-Superfund Cleanup Plan Falls Short." 2013. Indian Country Today Media Network, January 21. http://indiancountrytodaymedianetwork.com/2013/01/21/mohawks-say-epa-alcoa-superfund-cleanup-plan-falls-short-147131

"Molotov Cocktails and Guns Confiscated; Support Pours in for Mi'kmaq Protesters." 2013. Indian Country Today Media Network, October 18. http://indiancountrytodaymedianetwork.com/2013/10/18/support-pours-mikmaq-anti-fracking-protesters-new-brunswick-events-spring-and-sierra-club

Morgan, F. 2001. "Cultural Perspectives on Radiation: With Harmony They Were Placed." Workshop presented at "Creation of the Natural Order" Teachers' Seminar, Uranium Education Program, Diné College, Navajo Nation, July.

Morin, Brandi. 2013. "Federal and Provincial Governments Lose Appeal against First Nation Oil Sands Lawsuit." Indian Country Today Media Network, May 8. http://indiancountrytodaymedianetwork.com/2013/05/08/federal-and-provincial-governments-lose-appeal-against-first-nation-oil-sands-lawsuit

Murphy, Kim. 2001. "Front-Row Exposure to Global Warming; Climate: Engineers Say Alaskan Village Could Be Lost as Sea Encroaches." *Los Angeles Times*, July 8, A-1.

"NASA Scientist, Religious Leaders Arrested in Tar Sands Protest." 2011. Environment News Service, August 29. http://www.ens-newswire.com/ens/aug2011/2011-08-29-02.html

"NASA Study Finds Rising Arctic Storm Activity Sways Sea Ice, Climate." 2008. NASA Earth Observatory, October 6. http://earthobservatory.nasa.gov/Newsroom/NasaNews/2008/2008100627645.html

"National Parks Conservation Group Objects to Navajo Generating Station Proposal." 2013. Indian Country Today Media Network, November 13. http://indiancountrytodaymedianetwork.com/2013/11/13/national-parks-conservation-group-objects-navajo-generating-station-proposal-152238

"Native Alaskans Laud Environmental Protection Agency's Nixing of Pebble Mine in Bristol Bay." 2014. Indian Country Today Media Network, March 3. http://indiancountrytodaymedianetwork.com/2014/03/03/native-alaskans-laud-environmental-protection-agencys-nixing-pebble-mine-bristol-bay

Native Forest Network. 1994. "Quebec-Hydro Project May Destroy James Bay." Friends of the Earth, Victoria, British Columbia. http://jinx.sistm.unsw.edu.au/~greenlft/1994/137/137p14.htm

"Navajo Activists Spearhead Energy Transformation in the Four Corners Area." 2012. Indian Country Today Media Network, December 11. http://indiancountrytodaymedianetwork.com/article/navajo-activists-spearhead-energy-transformation-four-corners-area-146246

"Navajo Nation Could Get $1 Billion in Damages for Uranium Mess." 2013. Indian Country Today Media Network, December 18. http://indiancountrytodaymedianetwork.com/2013/12/18/navajo-nation-could-get-1-billion-damages-uranium-mess-152782

"Navajos Appeal to Inter-American Commission on Human Rights over Proposed Uranium Mining." 2011. Indian Country Today Media Network, May 16. http://indiancountrytodaymedianetwork.com/article/navajos-appeal-to-inter-american-commission-on-human-rights-over-proposed-uranium-mining-34304

"NCAI Condemns Keystone XL Pipeline." 2011. Indian Country Today Media Network, August 18. http://indiancountrytodaymedianetwork.com/article/ncai-condemns-keystone-xl-pipeline-48112

"The *New York Times* Profiles Blackfeet Drilling Dilemma." 2012. Indian Country Today Media Network, August 16. http://indiancountrytodaymedianetwork.com/article/%3Cem%3Ethe-new-york-times%3C/em%3E-profiles-blackfeet-drilling-dilemma-129595

"Nez Perce Victory: U.S. Forest Service Forbids Mega-Loads along Highway 12." 2013. Indian Country Today Media Network, September 20. http://indiancountrytodaymedianetwork.com/2013/09/20/nez-perce-victory-us-forest-service-forbids-mega-loads-along-highway-12-151372

Nikiforuk, Andrew. 1998. "Echoes of the Atomic Age: Cancer Kills Fourteen Aboriginal Uranium Workers." *Calgary Herald*, March 14, A-1, A-4. http://www.ccnr.org/deline_deaths.html

Nikiforuk, Andrew. 2014. "Large Dams of Mining Waste Leaking into Athabasca River: Study." Indian Country Today Media Network, February 26. http://indiancountrytodaymedianetwork.com/2014/02/26/large-dams-mining-waste-leaking-athabasca-river-study-153749?page=0%2C2

"19 Arrested in Idaho Protesting Mega-load Delivery." 2013. Indian Country Today Media Network, August 7. http://indiancountrytodaymedianetwork.com/2013/08/07/19-arrested-idaho-during-protest-against-megaload-delivery-150777

Norrell, Brenda. 2001. "Hopi and Navajo Protest Pipeline." *Indian Country Today Global Report* 129 (July 5–11). http://www.agrnews.org/issues/129/nationalnews.html

"Northern Quebec Cree Poisoned by Mining Pollutants: Report." 2001. Canadian Broadcasting Corporation News Online, October 20. http://cbc.ca/cgi-bin/view?/news/2001/10/19/cree_quebec011019

"Northwest Tribes and EPA Agree on Cleanup Plan for Upper Coeur d'Alene River Basin." 2012. Indian Country Today Media Network, September 4. http://indiancountrytodaymedianetwork.com/article/northwest-tribes-and-epa-agree-on-cleanup-plan-for-upper-coeur-d%E2%80%99alene-river-basin-132517

"Northwest Tribes Step Up Opposition to Proposed Coal Terminals." 2012. Indian Country Today Media Network, October 19. http://indiancountrytodaymedianetwork.com/article/northwest-tribes-step-up-opposition-to-proposed-coal-terminals-140853

"NPR Exposes Loophole That Allows Dumping of Toxic Oil Residue on Wind River Reservation." 2012. Indian Country Today Media Network, November 18. http://indiancountrytodaymedianetwork.com/article/npr-exposes-loophole-allows-dumping-toxic-oil-residue-wind-river-reservation-145738

"Nunavut Capital Battles Heat Wave." 2011. Indian Country Today Media Network, January 5. http://indiancountrytodaymedianetwork.com/article/nunavut-capital-battles-heat-wave-9443

"Obama Declares Bristol Bay Off Limits to New Oil and Gas Drilling Leases." 2014. Indian Country Today Media Network, December 17. http://indiancountrytodaymedianetwork.com/2014/12/17/obama-declares-bristol-bay-limits-new-oil-and-gas-drilling-leases-158329

"Once upon a Mine: National Institutes of Health Details Uranium's Toxic Legacy to Navajo." 2014. Indian Country Today Media Network, February 6. http://indiancountrytodaymedianetwork.com/2014/02/06/once-upon-mine-national-institutes-health-details-uraniums-toxic-legacy-navajo-153439

"Ottawa Blasts 'Radicals,' Blames 'Foreign Interests' for Northern Gateway Opposition." 2012. Indian Country Today Media Network, March 14. http://indiancountrytodaymedianetwork.com/article/ottawa-blasts-%27radicals%2C%27-blames-%27foreign-interests%27-for-northern-gateway-opposition

Overpeck, J. T., M. Strum, J. A. Francis, D. K. Perovich, M. C. Serreze, R. Benner, E. C. Carmack. F. S. Chapin III, S. C. Gerlach, L. C. Hamilton, L. D. Hinzman, M. Holland, H. P. Huntington, J. R. Key, A. H. Lloyd, G. M. MacDonald, J. McFadden, D. Noone, T. D. Prowse, P. Schlosser, and C. Vorosmarty. 2005. "Arctic System on Trajectory to New, Seasonally Ice-Free State." *EOS: Transactions of the American Geophysical Union* 86, no. 34 (August 23): 309, 312.

Pasternak, Judy. 2010. *Yellow Dirt: An American Story of a Poisoned Lands and a People Betrayed.* New York: Free Press.

"PBS News Hour: Lummi Fighting Cherry Point Coal Export Terminal." 2012. Indian Country Today Media Network, August 4. http://indiancountrytodaymedianetwork.com/2013/08/04/pbs-newshour-highlights-lummi-battle-over-cherry-point-coal-export-terminal-150726

PCB Working Group. N.d. "Communities Respond to PCB Contamination." http://www.ipen.org/circumpolar2.html

"PD 2000 Projects." 2001. Arctic Monitoring and Research—Project Directory, April 11. http://amap.no/pd2000.htm

Pegg, J. R. 2004. "The Earth Is Melting, Arctic Native Leader Warns." Environment News Service, September 16. http://www.ens-newswire.com/ens/sep2004/2004-09-16-10.asp

Pember, Mary Annette. 2012. "Climate Change Threatens the Ojibwe's Wild Rice Harvest." Indian Country Today Media Network, November 2. http://indiancountrytodaymedianetwork.com/article/climate-change-threatens-the-ojibwes-wild-rice-harvest-143579

Pember, Mary Annette. 2013a. "Automatic Weapons & Guards in Camo: Welcome to Mining Country, Wis." Indian Country Today Media Network, July 9. http://indiancountrytodaymedianetwork.com/2013/07/09/automatic-weapons-guards-camo-welcome-mining-country-wis-150342

Pember, Mary Annette. 2013b. "County Declares Harvest Camp Illegal, Moves to Shut It Down." Indian Country Today Media Network, July 23. http://indiancountrytodaymedianetwork.com/story/wis-mining-war

Pember, Mary Annette. 2013c. "Racism & Violent Threats: Wis. Mining War Gets Uglier and Scarier." Indian Country Today Media Network, July 16. https://indiancountrytodaymedianetwork.com/2013/07/16/wis-mining-racism-and-threats-reach-facebook-150449

Pember, Mary Annette. 2013d. "Republican Favoritism of GTAC Gets Public Comment Session." Indian Country Today Media Network, September 3. http://indiancountrytodaymedianetwork.com/2013/09/03/republican-favoritism-gtac-gets-public-comment-session-151137

Pember, Mary Annette. 2013e. "Who Is Illegal in Wis. Mining Country? Harvest Camp or GTAC?" Indian Country Today Media Network, July 20. http://indiancountrytodaymedianetwork.com/2013/07/20/who-illegal-wis-mining-country-harvest-camp-or-gtac-150512

Pember, Mary Annette. 2015. "A Bad River Win; Gogebic Taconite Putting Wisc. Mine on Hold." Indian Country Today Media Network, March 3. http://indiancountrytodaymedianetwork.com/2015/03/03/bad-river-win-gogebic-taconite-putting-wisc-mine-hold-159434

Peterson, Diane J. 2000."Two Twin Cities Churches Protest Hydro Power Injustices." *Earthkeeping News* 9, no. 5 (July/August). http://www.nacce.org/2000/manitoba.html

Phillip, Chief Stewart. 2001. "The Criminalization of Skwelkwek'welt Defenders." September 4. https://groups.yahoo.com/neo/groups/sovernspeakout/conversations/topics/5722

Pierpoint, Mary. 2001. "Reno-Sparks Fights Kitty Litter Mine; High Traffic, Dust, Potentially Toxic Minerals All Concerns." *Indian Country Today*, December 12, A-1.

Pogash, Carol. 2013. "Governor, Chippewas Battle over Mine." *USA Today*, September 9, 4-A.

"Pour Him a Tall One! Mining Exec Insists He'd Drink Water from Tailings Pond." 2014. Indian Country Today Media Network, August 7. http://indiancountrytodaymedianetwork.com/2014/08/07/pour-him-tall-one-mining-exec-insists-hed-drink-water-tailings-pond-156294

"Prairie Island Indian Community Calls for Permanent Nuclear-Waste Solution." 2012. Indian Country Today Media Network, September 5. http://indiancountrytodaymedianetwork.com/article/prairie-island-indian-community-calls-for-permanent-nuclear-waste-solution-132820

"Protecting Knowledge: Native Youth Movement Protest at Sun Peaks Resort; Interview with Amanda Soper, Spokesperson for the Native Youth Movement, and Chris Rogers, Spokesperson for Sun Peaks Resort." 2001. CBC Radio, B.C. Almanac, August 24.

Pugliese, David. 2001. "An Expensive Farewell to Arms: The U.S. Has Abandoned 51 Military Sites in Canada. Many Are Polluted, and Taxpayers Are Paying Most of the $720 Million Cleanup Cost." *Montreal Gazette*, April 28, B-1.

Purdom, Tim. 2001. "Radiation Victims Rally; Marchers Complain about IOUs, Delays in Payments." *Gallup Independent*, March 13. http://www.cia-g.com/~gallpind/3-13-01.html#anchor1

Qiu, Jane. 2009. "Arctic Ecology: Tundra's Burning." *Nature* 461 (September 3): 34–36.

"Quapaw Take Lead in Tar Creek Superfund Site Cleanup in Oklahoma, Kansas." 2014. Indian Country Today Media Network, June 4. http://indiancountrytodaymedianetwork.com/2014/06/04/quapaw-take-lead-tar-creek-superfund-site-cleanup-oklahoma-kansas-155155

"Quebec Admits Toxic Waste Poisons Cree." 2001. Canadian Broadcasting Corporation News Online, October 26. http://cbc.ca/cgi-bin/view?/news/2001/10/26/cree_tox011026

"Quebec Will Shelve Huge Hydroelectric Project Indefinitely." 1994. *Omaha World-Herald*, November 19, 7.

Raloff, J. 1998. "Picturing Pesticides' Impacts on Kids." *Science News* 153, no. 23 (June 6): 358.

"Rare Sighting of Wasp North of Arctic Circle Puzzles Residents." 2004. Canadian Broadcasting Corporation, September 9. http://www.cbc.ca/story/science/national/2004/09/09/wasp040909.html

Rauber, Paul. N.d. "On Top of the World: Midsummer on Baffin Island: Where the Sun Never Sets on the Polar Bear's Empire and Staying Alive Is a Round-the-Clock Job." Sierra Club. http://www.sierraclub.org/sierra/199803/baffin.html

Reese, April. 2011. "Navajo Group to Take Uranium Mine Challenge to Human Rights Commission." *New York Times*, May 11. http://www.nytimes.com/gwire/2011/05/12/12greenwire-navajo-group-to-take-uranium-mine-challenge-to-33718.html

"Resisting Destruction: Chronology of the Lubicon Crees' Struggle to Survive." N.d. Lubicon Solidarity. http://www.lubiconsolidarity.ca/resisting.html

Riordan, Michael. 2014. "Don't Sell Cheap U.S. Coal to Asia." *New York Times*, February 13, A-23. http://www.nytimes.com/2014/02/13/opinion/dont-sell-cheap-us-coal-to-asia.html

Rolfe, Christopher. 1996. "Comments on the British Columbia Greenhouse Gas Action Plan." West Coast Environmental Law Association presentation to the Air and Water Management Association, April 17. http://www.wcel.org/wcelpub/11026.html

Roslin, Alex. 2000. "Crees Revive Hydro Project." *Montreal Gazette*, January 21. http://www.montrealgazette.com/news/pages/010121/5036705.html

Ross, John. 1996. "Is Zapatista Rebellion Rooted in Oil?" *Earth Island Journal* 11, no. 2: 20.

Russo, Kurt. 2012. "Review of *Asserting Native Resilience: Pacific Rim Indigenous Nations Face the Climate Crisis*." *American Indian Culture and Research Journal* 37, no. 2: 234.

Schiermeier, Quirin. 2007. "On Thin Ice: The Arctic Is the Bellwether of Climate Change." *Nature* 441 (May 10): 146–47.

Schilling, Vincent. 2011. "Documentary Featuring the Fight between Ford and Ramapough Indians Premieres Tonight." Indian Country Today Media Network, July 18.http://indiancountrytodaymedianetwork.com/article/documentary-featuring-the-fight-between-ford-and-ramapough-indians-premieres-tonight-43261

Schilling, Vincent. 2012. "Pine Ridge Residents Halt Canadian Mine Equipment Transportation through Reservation." Indian Country Today Media Network, March 9. http://indiancountrytodaymedianetwork.com/article/pine-ridge-residents-halt-canadian-mine-equipment-transportation-through-reservation-102246

Schilling, Vincent. 2013a. "Behind the Front Lines of the Elsipogtog Battle over Fracking." Indian Country Media Network, November 20. http://indiancountrytodaymedianetwork.com/2013/11/20/behind-lines-elsipogtog-aptn-journalist-ossie-michelin-gives-backstory-152354

Schilling, Vincent. 2013b. "Woman with Eagle Feather: The Photo 'Heard' round the World." Indian Country Today Media Network, November 21. http://indiancountrytodaymedianetwork.com/2013/11/21/woman-eagle-feather-photo-heard-round-world-152357

Schneider, Andrew. 2001. "'A Wounded Mountain Spewing Poison': The Mining of the West: Profit and Pollution on Public Lands: A *Post-Intelligencer* Special Report."

*Seattle Post-Intelligencer*, June 12. http://seattlepi.nwsource.com/specials/mining/27076_lodgepole12.shtml

Schneider, Iyawata Britten. N.d. "Environmental and Human Rights Devastation in the Southwest." *Country Road News*. http://www.countryroadchronicles.com/Articles/CountryRoadNews/devastation.html

Schneider, Keith. 1990. "Washington Nuclear Plant Poses Risk for Indians." *New York Times*, September 3, A-9.

"Scientists Look to 'Traditional Knowledge' to Help Understand Climate Change." 2001. Canadian Broadcasting Corporation News, March 22. http://cbc.ca/cgi-bin/view?/news/2001/03/21/climate010321

Screen, James A., and Ian Simmonds. 2010. "The Central Role of Diminishing Sea Ice in Recent Arctic Temperature Amplification." *Nature* 464 (April 29): 1334–37.

Seldon, Ron. 2002. "Families Sicken and Die in Mold-Plagued HUD Housing; Residents Fight Back against Substandard Construction." *Indian Country Today*, April 3, B-1, B-2.

Shaffer, David. 2012. "Prairie Island Tribe Seeks Action on Nuclear Waste." *Star-Tribune*, September 3. http://www.startribune.com/local/168337166.html

Shogren, Elizabeth. 2012. "Loophole Lets Toxic Oil Water Flow over Indian Land." KERA: News for North Texas, November 15. http://keranews.org/post/loophole-lets-toxic-oil-water-flow-over-indian-land

Sierra Club. N.d. "Hudson Bay/James Bay Watershed Eco-Region." http://www.sierraclub.org/ecoregions/hudsonbay.asp

"Silver Valley Cleanup Plan to Cost $635 Million." 2012. *Seattle Times*, August 28. http://seattletimes.com/html/localnews/2019011322_apidsilvervalleycleanup1stldwritethru.html?syndication=rss

Simmonds, Ian, and Kevin Keay. 2009. "Extraordinary September Arctic Sea Ice Reductions and Their Relationships with Storm Behavior over 1979–2008." *Geophysical Research Letters* 36 (October 14): L19715. doi:10.1029/2009GL039810

Smith, Mitch. 2015. "Grass-Roots Push in the Plains to Block the Keystone Pipeline's Path." *New York Times*, May 6. http://www.nytimes.com/2015/05/06/us/grass-roots-push-in-the-plains-to-block-the-keystone-pipelines-path.html

Smitten/Vimeo, Susan. 2014. "First Nations Exult as Canadian Government Rejects Taseko's New Prosperity Mine." Indian Country Today Media Network, February 27. http://indiancountrytodaymedianetwork.com/2014/02/27/first-nations-exult-canadian-government-rejects-tasekos-new-prosperity-mine-153777

Sockbeson, Rebecca. 1999. "Statement by Rebecca Sockbeson for IRATE (Indigenous Resistance Against Tribal Extinction), and I.E.N. (Indigenous Environmental Network)." International POPs Elimination Network, PCB Working Group, September 8. http://www.ipen.org/irate.html

Sontag, Deborah, and Brent McDonald. 2014. "In North Dakota, a Tale of Oil, Corruption and Death." *New York Times*, December 29. http://www.nytimes.com/2014/12/29/us/in-north-dakota-where-oil-corruption-and-bodies-surface.html

Stone, Laura. 2013. "Chief Atleo Slams RCMP for 'Extreme Use of State Force' in N.B. Shale Gas Protest." Global News (Canada), October 18. http://globalnews.ca/news/910934/chief-atleo-slams-rcmp-for-extreme-use-of-state-force-in-n-b-shale-gas-protest/?utm_source=facebook-twitter&utm_medium=link&utm_campaign=community

Sudetic, Chuck. 1999. "As the World Burns." *Rolling Stone*, September 2, 97–106, 129.

"Sustainable Development in the Hudson Bay: James Bay Bio-Region Canadian Arctic Resources Committee." N.d. Environmental Committee of Sanikiluaq, Rawson Academy of Aquatic Science.http://www.carc.org/pubs/v19no3/2.htm

Suzuki, David. 2000. "Science Matters: POP Agreement Needed to Eliminate Toxic Chemicals." December 6. http://www.davidsuzuki.org/Dr_David_Suzuki/Article_Archives/weekly12060002.asp

"Tar Sands Opposition Unites Burnaby Leaders, First Nations, as 60 Arrested in B.C. Protests." 2014. Indian Country Today Media Network, November 26. http://indiancountrytodaymedianetwork.com/2014/11/26/tar-sands-opposition-unites-burnaby-leaders-first-nations-60-arrested-bc-protests-158031

Taylor, Kevin. 2013. "Nez Perce Leaders Stand Firm on Frontlines of Mega-Load Transport." Indian Country Today Media Network, August 9. http://indiancountrytodaymedianetwork.com/2013/08/09/nez-perce-leaders-stand-firm-frontlines-mega-load-transport-150809

Taylor, Robert. 2002. "Cree Leaders Jailed in Protest over Power Deal." *Indian Country Today*, February 17. http://www.indiancountry.com/?1013955409

"Thinning Ice More Fragile and Mobile." 2012. *Nature* 491 (November 15): 304.

Thomas, William. 1998. "Salmon Dying in Hot Waters." Environment News Service, September 22. http://www.econet.apc.org/igc/en/hl/9809244985/hl11.html

"360 Scientists Urge Environmental Protection Agency to Quash Bristol Bay Pebble Mine." 2014. Indian Country Today Media Network, February 14. http://indiancountrytodaymedianetwork.com/2014/02/14/360-scientists-urge-environmental-protection-agency-quash-bristol-bay-pebble-mine-153559

"Thunderstorms Are Latest Evidence of Climate Change." 2000. Associated Press Canada, November 15. http://abcnews.go.com/sections/science/DailyNews/arctic_thunder001115.html

"Time to Act on Nunavut Climate Change." 2001. Canadian Broadcasting Corporation, March 16. http://north.cbc.ca/cgi-bin/templates/view.cgi?/news/2001/03/16/16nunmoose

Tingley, Martin P., and Peter Huybers.2013. "Recent Temperature Extremes at High Northern Latitudes Unprecedented in the Past 600 Years." *Nature* 496 (April 11): 201–5.

Toensing, Gale Courey. 2013. "Anaya Seeks Information for Study on Extractive and Energy Industries." Indian Country Today Media Network, February 6. http://indiancountrytodaymedianetwork.com/2013/02/06/anaya-seeks-information-study-extractive-and-energy-industries-147495

Tollefson, Jeff. 2013. "Climate Science: A Line in the Sands." *Nature* 500 (August 8): 136–37. http://www.nature.com/news/climate-science-a-line-in-the-sands-1.135150

Tomsho, Rupert. 1990. "Reservations Bear the Brunt of New Pollution." *Wall Street Journal*, November 29, 1.

Toynbee, Arnold. 1973. "The Genesis of Pollution." *New York Times*, September 16, n.p.

Trahant, Mark. 2012. "Elections 2012: Vote for Coal? Northwest Tribes Raise New Concerns." Indian Country Today Media Network, October 1. http://indiancountrytodaymedianetwork.com/article/elections-2012%3A-vote-for-coal%3F-northwest-tribes-raise-new-concerns-137150

"Treaty Victory as Northwest Tribes Celebrate Oregon Coal Train Rejection." 2014. Indian Country Today Media Network, August 21. http://indiancountrytodaymedianetwork.com/2014/08/21/treaty-victory-northwest-tribes-celebrate-oregon-coal-train-rejection-156520

"Tribal Members Sign Treaty Calling for an End to Alberta Oil Sands Development and Keystone XL." 2013. Indian Country Today Media Network, January 31. http://indiancountrytodaymedianetwork.com/2013/01/31/tribal-members-sign-treaty-calling-end-alberta-oil-sands-development-and-keystone-xl

Troian, Martha. 2013a. "First Nation Moves to Evict Fracking Co. from Lands Held in Trust." Indian Country Today Media Network, October 3. http://indiancountrytodaymedianetwork.com/2013/10/03/elsipogtog-first-nation-wrests-lands-fracking-co-mikmaq-treaty-day-151564

Troian, Martha. 2013b. "Police in Riot Gear Tear-Gas and Shoot Mi'kmaq Protesting Gas Exploration in New Brunswick." Indian Country Today Media Network, October 17. http://indiancountrytodaymedianetwork.com/2013/10/17/police-riot-gear-tear-gas-and-shoot-mikmaq-protesting-gas-exploration-new-brunswick

Troian, Martha. 2014. "Court Tells Lubicon Lake Cree in Alberta to Stop Blocking Fracking Activities." Indian Country Today Media Network, February 6. http://indiancountrytodaymedianetwork.com/2014/02/06/court-tells-lubicon-lake-cree-alberta-stop-blocking-fracking-activities-153435

"Tsilhqot'in Acquiesce to Test Drilling, Still Determined to Stop New Prosperity Mine." 2012. Indian Country Today Media Network, March 2. http://indiancountrytodaymedianetwork.com/article/tsilhqotin-acquiesce-to-test-drilling-still-determined-to-stop-new-prosperity-mine-101023

"Turtle Mountain Tribal Council Bans Fracking." 2012. Indian Country Today Media Network, November 27. http://indiancountrytodaymedianetwork.com/article/turtle-mountain-tribal-council-bans-fracking-64866

"Unions Back Lubicons." 2000. Indigenous Environmental Network, August 16. http://www.ienearth.org/lubicon.html#canada

"Uranium Miners, Families Bring Tales of Pain to Washington." 2000. *Arizona Republic*, April 15. http://www.ienearth.org/mining_campaign_1c.html

"Uranium Mining and the Laguna People." 1996. Dorothy Purley, interviewed by Susan Lee, July 1995, in Paguate Village on the Laguna Pueblo Reservation. *Synthesis/Regeneration* 10 (Spring). http://www.greens.org/s-r/10/10-07.html

"Uranium/Nuclear Issues and Native Communities." 2001. Indigenous Environmental Network.http://www.ienearth.org/nuciss.html

Valencia, Angel. 2000. "Statement of Angel Valencia, Yoemem Tekia Foundation, Tucson, Arizona POPs Negotiations, March 23, 2000." Native News: The Mail Archive, March 27. http://www.mail-archive.com/natnews@onelist.com/msg00572.html

Vance, Chris, and John Shafer. 1995. "Voices of Resistance." Radio transcript, CFUV 102 FM, (community radio station, University of Victoria, British Columbia), transcribed in 1995 and reprinted in the magazine *All That's Left Is Struggle*. http://www.finearts.uvic.ca/~vipirg/SISIS/sov/allcree.html

Vandevelder, Paul. 2000. "Between a Rock and a Dry Place." *Native Americas* 17, no. 2 (Summer): 10–15.

Vandevelder, Paul. 2001. "A Native Sense of Earth." *Native Americas* 18, no. 1 (Spring): 42–49.

Walker, Richard. 2014. "Lummi Totem Pole Journey Rallies Voices against Environmental Destruction." Indian Country Today Media Network, September 2. http://indiancountrytodaymedianetwork.com/2014/09/02/lummi-totem-pole-journey-rallies-voices-against-environmental-destruction-156696

Walker, Richard. 2015. "Lummi Call Coal Terminals an Absolute No-Go, Invoking Treaty Rights." Indian Country Today Media Network, February 26. http://indian

countrytodaymedianetwork.com/2015/02/26/lummi-call-coal-terminals-absolute-no-go-invoking-treaty-rights-159381

Wanamaker, Tom. 2002. "Hopi Tribe Opposes Lease Extension on Water-Thirsty Black Mesa Mine." *Indian Country Today*, February 13, C-1.

"Warming Past, Present." 2009. *Omaha World-Herald*, September 4, 5-A.

Warrick, Joby. 2015. "Gold vs. Salmon: EPA Becomes Target by Employing Rare Preemptive 'Veto.'" *Washington Post*, February 15. http://www.washingtonpost.com/national/health-science/internal-memos-spur-accusations-of-bias-as-epa-moves-to-block-gold-mine/2015/02/15/3ff101c0-b2ba-11e4-854b-a38d13486ba1_story.html

"Watershed Heroes: Colville Confederated Tribes Win Sierra Club Award for Battling British Columbia Smelter." 2013. Indian Country Today Media Network, March 6. http://indiancountrytodaymedianetwork.com/2013/03/06/watershed-heroes-colville-confederated-tribes-win-sierra-club-award-battling-british

Watt-Cloutier, Sheila. 2003. "Speech to Conference of Parties to the United Nations Framework Convention on Climate Change," Milan, Italy, December 10.

Weaver, Jace, ed. 1996. *Defending Mother Earth: Native American Perspectives on Environmental Justice.* New York: Maryknoll.

Weaver, Kasey. 2001. "Sacred Sites Threatened." *We Have Many Voices* 1, no. 13 (July 21). http://www.turtletrack.org/ManyVoices/Issue_13/SacredSites_721.htm

Weinberg, Bill. 1994. "Flooding the Jungle." *Native Americas* 11, no. 2 (Summer): 43.

"Welcome to Fearless Summer: Protesters Block Keystone XL Construction." 2013. Indian Country Today Media Network, June 24. http://indiancountrytodaymedianetwork.com/2013/06/28/fracking-troubles-atlantic-first-nations-after-two-dozen-protesters-arrested-150192

Wilkin, Dwane. 1997. "Global Warming Poses Big Threats to Canada's Arctic." *Nunatsiaq News*, November 21. http://www.nunatsiaq.com/archives/back-issues/71121.html#6

Wilson, Reid. 2014. "Jewell 'Profoundly' Disappointed by Land Exchange at Sacred Native American Sites." *Washington Post*, December 8. http://www.washingtonpost.com/blogs/govbeat/wp/2014/12/06/jewell-profoundly-disappointed-by-land-exchange-at-sacred-native-american-sites

Windeyer, Chris. 2011. "South Baffin Swelters in Winter Heat Wave; 'It Doesn't Show Any Signs of Abating.'" *Nunatsiaq News*, January 4. http://www.nunatsiaqonline.ca/stories/article/98789_south_baffin_swelters_in_winter_heat_wave

Wohlforth, Charles. 2004. *The Whale and the Supercomputer: On the Northern Front of Climate Change*. New York: North Point Press/Farrar, Straus and Giroux.

Woodard, Stephanie. 2012. "Moapa Paiutes Find Solar Solution Amid Coal Ash Plague." Indian Country Today Media Network, August 17. http://indiancountrytodaymedianetwork.com/article/moapa-paiutes-find-solar-solution-amid-coal-ash-plague-129554

Woodard, Stephanie. 2013. "Planned Oil Pipeline Must Cross Pine Ridge's Water-Delivery System." Indian Country Today Media Network, September 21. http://indiancountrytodaymedianetwork.com/article/planned-oil-pipeline-must-cross-pine-ridge%25e2%2580%2599s-water-delivery-system-54942.html

Woody, R. L., B. Jack, and V. Bizahaloni. 1981. *Social Impact of the Uranium Industry on Two Navajo Communities*. Tsaile, AZ: Navajo Community College.

"Wyoming Groundwater Again Tests Positive for Fracking-Related Chemicals on Wind River Reservation." 2012. Indian Country Today Media Network, September 27.

http://indiancountrytodaymedianetwork.com/article/wyoming-groundwater-again-tests-positive-for-fracking-related-chemicals-on-wind-river-reservation-136259

York, Geoffrey. 2001. "Russian City Ravaging Arctic Land." *Toronto Globe and Mail*, July 25, A-1, A-11.

"Yup'ik Villages Ravaged by Fierce Alaska Storms." 2013. Indian Country Media Network, November 20. http://indiancountrytodaymedianetwork.com/2013/11/20/yupik-villages-ravaged-fierce-alaska-storms-152341

Zhorov, Irina. 2012a. "Extra Testing Being Conducted for Wind River Reservation Uranium Contamination." Indian Country Today Media Network, August 29. http://indiancountrytodaymedianetwork.com/2012/08/29/extra-testing-being-conducted-wind-river-reservation-uranium-contamination-131575

Zhorov, Irina. 2012b. "Reopening Silver Mine Could Save Town, but Destroy the Sacred Site of the Huicholes." Indian Country Today Media Network, August 7. http://indiancountrytodaymedianetwork.com/article/reopening-silver-mine-could-save-town,-but-destroy-the-sacred-site-of-the-huicholes-127755

# Index

**About the Author**

BRUCE E. JOHANSEN is Jacob J. Isaacson Professor of Communication and Native American Studies at the University of Nebraska at Omaha, where he has been teaching and writing since 1982. He has authored 42 published books. Johansen writes frequently about environmental subjects, including *The Encyclopedia of Global Warming Science and Technology* (2 vols., 2009), *Global Warming in the 21st Century* (3 vols., 2006), *The Global Warming Desk Reference* (2001), *The Dirty Dozen: Toxic Chemicals and the Earth's Future* (2003), and *Indigenous Peoples and Environmental Issues* (2004), a 200,000-word encyclopedia of indigenous peoples' struggles with corporations with a worldwide scope. He also writes as a journalist in several national forums, including the *Washington Post* and *The Progressive*, with letters to the editor in *The Atlantic*, the *New York Times*, *National Geographic*, the *Wall Street Journal*, and others. He was coeditor of the *Encyclopedia of American Indian History*, a four-volume set (ABC-CLIO, 2007), as well as a two-volume *Praeger Handbook of Contemporary Native American Issues* (2007). He also wrote *The Global Warming Combat Manual: Serious Solutions for a Sustainable World* (Praeger, 2008). His *Eco-Hustle! Global Warming, Greenwashing, and Sustainability* was published in 2015.

www.ingramcontent.com/pod-product-compliance
Ingram Content Group UK Ltd.
Pitfield, Milton Keynes, MK11 3LW, UK
UKHW020844290926
15470UKWH00004B/1541

9 781440 831